SEISMIC LOVE WAVES

POVERKHNOSTNYE VOLNY LYAVA

ПОВЕРХНОСТНЫЕ ВОЛНЫ ЛЯВА

SEISMIC LOVE WAVES

Z. S. Andrianova
Institute of Chemical Physics

V. I. Keilis-Borok
Institute of Physics of the Earth

A. L. Levshin
Institute of Physics of the Earth

M. G. Neigauz
Institute of Chemical Physics

Academy of Sciences of the USSR

Translated from Russian by
F. M. C. Goodspeed
Department of Mathematics
Laval University
Quebec, Canada

CONSULTANTS BUREAU · NEW YORK · 1967

About the Authors:

Zoya Sidorovna Andrianova was born in 1938. She was graduated from Moscow University in 1960 and is now at the Institute of Chemical Physics of the Academy of Sciences of the USSR. Andrianova is the author of several articles on the numerical solution of differential equations.

Vladimir Isaakovich Keilis-Borok was born in 1921 and was graduated from the Institute of Geological Exploration in 1943. He received his doctorate in physical mathematics in 1953. Director of the interpretation-methods section of the Institute of Physics of the Earth of the Academy of Sciences of the USSR, he is the author of the monograph "Interference Surface Waves." His principal work concerns the problems of elastic-wave theory, the machine interpretation of seismic observations, and the detection of nuclear explosions.

Anatolii L'vovich Levshin was born in 1932 and was graduated from Moscow University in 1954. He is now at the Institute of Physics of the Earth of the Academy of Sciences of the USSR studying the theory and interpretation of surface waves in engineering seismic exploration and seismology.

Militsa Genrikhovna Neigauz was born in 1929, and was graduated from Moscow University. She is now at the Institute of Chemical Physics of the Academy of Sciences of the USSR and is associated with the mathematical school of I. M. Gel'fand. She is the author of several articles on the spectral theory of operators, optimum control, and stability theory.

The Russian text, originally published by Nauka Press in Moscow in 1965 for the Institute of Physics of the Earth and the Institute of Chemical Physics of the Academy of Sciences of the USSR, has been extensively corrected and expanded by the authors for the English edition.

З. С. Андрианова, В. И. Кейлис-Борок, А. Л. Левшин, М. Г. Нейгауз

Поверхностные волны Лява

Library of Congress Catalog Card No. 66-19935

ISBN 978-1-4684-8606-3   ISBN 978-1-4684-8604-9 (eBook)
DOI 10.1007/978-1-4684-8604-9

© *1967 Consultants Bureau*

Softcover reprint of the hardcover 1st edition 1967

A Division of Plenum Publishing Corporation
227 West 17 Street, New York, N. Y. 10011
All rights reserved

FOREWORD TO THE ENGLISH EDITION

Despite the fact that there have been a great many studies of Love waves, the present book is not a summary of previous work; it contains new practical results and is completely dedicated to questions of method. This requires an explanation of why we should consider methods of investigation which are already very effective. The explanation is related to the development of the problem.

The electronic computer and new recording apparatus have recently and suddenly revealed great new possibilities in the investigation of surface waves. The new potentialities naturally first developed in conjunction with a method which, in reference to petroleum resources in the USA, is called wildcatting. Here investigators try to obtain specific results without any interest in the optimum nature of the method, in the uniqueness of the results, in the investigation of allied problems, etc.

We have a high regard for the useful results obtained and the attempts made during this period. It is clear, however, that we have arrived at a stage in which better planned and more complex investigations must be carried out. At this time we need a summary of the general fundamental aspects of the problem. We refer to the general theoretical properties of Love waves, to methods for calculating the parameters of Love waves, to the relations between the properties of Love waves and various features of the medium and the source, to the accuracy of the observations needed for the determination of various details of profiles of the crust or mantle, to the uniqueness of the results of interpretation, etc.

These general problems are considered in the present book. A short summary of the contents is given below in the introduction. The text has been completely revised and an appendix added for the present edition; the main additions to the original text are devoted to a consideration of the influence of sphericity and absorption and an analysis of theoretical seismology. The appendix was written with the participation of E. V. Vil'kovich and N. P. Grudeva.

We believe that this book will be of interest to English-speaking readers who are interested in Love waves, even though numerous articles on this subject have been published in the English technical literature.

The Authors

PREFACE

Among seismic waves generated by widely different types of sources, the waves that are the most interesting and of the longest duration are usually surface waves (which are also called normal waves, interference and channel waves, natural vibrations of layers, etc.). These waves are distinguished by their dispersion and resonance, and are used in many investigations — in the determination of the structure of the medium (the existence of surface layers and waveguides), in the determination of the coordinates and properties of the source (in particular its energy and mechanism), in the identification of subterranean explosions, in the mapping of microseisms, in the tracing of storms, etc. Outlines of the wide literature concerning surface waves can be found in [2, 9, 40].

Surface waves can be classed as Rayleigh or Love waves, depending on their velocity and polarization: the displacements of particles in Rayleigh waves are parallel, and the displacements in Love waves perpendicular, to the vertical plane containing the direction of propagation (some variation from this polarization is possible close to the source or in the presence of horizontal inhomogeneities in the medium).

The present work is devoted to the calculation and investigation of Love waves in a vertically layered medium. The method used is based on the spectral theory of linear differential operators.* This method is much more effective than previously applied methods based on the approximation of the medium by a set of homogeneous layers.

Chapter I describes the theory of Love waves. In it we make the following assumptions: The medium is a perfectly elastic, isotropic half-space, the parameters of which are piecewise continuous functions of the depth, independent of the horizontal components. The contact between particles on the planes of discontinuity is assumed to be perfect. Beginning at a certain depth the parameters of the medium become constant and the volume-wave velocity reaches its maximum. In the source, the volume forces are distributed symmetrically relative to a vertical axis. The dependence of the force in the source on the time and on the horizontal coordinates is assumed to be sufficiently smooth (the conditions imposed on the source can be somewhat relaxed but this is not physically essential).

In Section 1 we give the original formulas. By using the ordinary method of separation of variables and some extra calculations, we express the nonstationary displacements in Love waves in a three-dimensional medium (and in particular their asymptotic representation at large horizontal distances from the source) in terms of the characteristic functions of a Sturm-Liouville operator.

In Section 2 we consider a method for the calculation and investigation of Love waves. This method yields not only the dispersion, but also the intensity — in practice for any mode and any velocity and density profile in a wide range of frequencies. A program is written for using this method with an electronic digital computer.

---

*This theory has previously been used only in the qualitative investigation of the dispersion of Love waves in an inhomogeneous layer on an absolutely rigid half-space [43].

In Section 3 we consider the fundamental theoretical properties of Love waves:* their dispersion; the dependence of their intensity on the vibration frequency, the source depth, and the mode number; the nature of nonstationary vibrations and methods of constructing theoretical seismograms; the laws of similarity and the reciprocity principle. We describe in detail the specific properties of vibrations in the presence of waveguides. In conclusion we give the basic information needed for the application of the program that has been built up.

Chapters II and III contain a discussion on the possibility of using Love waves for the investigation of the structure of the crust and the upper mantle; it is explained what observations are necessary for the determination of certain profile elements (waveguides, zones of higher gradient, boundaries between layers), and what details of a profile can be determined from given observations (one or several of the layer modes). The two chapters are similarly planned.

Sections 4 and 7 contain descriptions of frequency characteristics; from them we can determine the range of periods and focal depths for which the first three or four modes are sufficiently intense. The resulting frequency characteristics are used in deciphering the wave picture, in obtaining source depths, and in obtaining more accurate earthquake magnitudes; they can also be used in the analysis of microseisms, the investigation of resonance properties of layers, etc.

Calculations show a very slight decrease in intensity with increasing mode number; in some examples there is almost no variation in the intensity of the first forty modes. We have not yet succeeded in establishing whether real records are actually formed by the superposition of such a great number of modes. We therefore confine ourselves to the consideration of those time intervals and vibration periods for which only the first 3-4 modes are recorded.

In Sections 5 and 8 we compare the phase- and group-velocity dispersion of these modes for various velocity profiles. We explain which sections of the dispersion curves are needed for the determination of those elements of a profile which are of direct interest in seismology. The basic lack of definiteness in several common methods of interpretation is demonstrated. Thus from the first (fundamental) mode we can determine for the earth's crust only a certain approximately linear combination of thickness and mean velocity, and this only when the thicknesses and the mean velocities of the layers covering the strictly crystalline crust are known; for the upper mantle we can determine only the mean velocity over the large depth intervals (of the order of hundreds of kilometers), while waveguides can remain undetected if their parameters are close to those which can be received at present.

Methods are described for overcoming this lack of definiteness and for determining the basic details of a profile. These methods reduce to the simultaneous employment of several modes (appropriate to the frequency intervals employed) or of body and surface waves. However, for any conceivable system of observations, interpretation must be made in terms of a set of profiles, all of which may satisfy the data, rather than a single, unique profile.

The best possible prospect of determining waveguides is clearly in the use of specific vibrations related to a secondary group-velocity minimum of the third mode.

In Section 8 we compare some calculated results with experimental observations. An example is given of the simultaneous interpretation of surface and body waves; a group of profiles is found compatible with the observed dispersion of the first mode of Love waves in the travel time of the S Wave.

In Sections 6 and 9 we clarify the conditions for obtaining data required for the solution of the problems considered in Sections 5 and 8. Theoretical seismograms are obtained for nonstationary Love waves. Criteria are given for the separation of the different modes.

Chapter IV contains a discussion of the relation between the nature of the source and the observations of Love waves. The possibility is considered of estimating the focal depth (when the focus is in the crust) from

---

*Readers who are not interested in the purely mathematical side of the problem can begin with Section 3, which starts with the basic definitions.

the spectrum of the fundamental mode (Section 10); corrections of the magnitude for focal depth are considered in Section 11; the relation between the azimuthal distribution of intensity and the mechanism of an earthquake is discussed in Section 12.

This summary of the contents of the book has left the authors with a feeling of some dissatisfaction. If we were rewriting the book we would probably have the models of the medium in Sections 5 and 8 vary automatically (i.e., by using the appropriate program) as was done in the solution of the inverse problem ([28, 32], Paragraph 2, Section 8 of this book), and would investigate the derivatives of the phase and group velocity with respect to the profile parameters. We would also investigate the quantitative correlation between profiles and group-velocity curves for higher modes.

We decided, however, not to pospone the publication of the book for consideration of the above questions in the hope that it can, in its present form, at least partially show the great possibilities inherent in the use of Love waves for seismology.

In the preparation of the mathematical part of the book we had the valuable advice of Professor V. B. Lidskii, Professor A. Ya. Povzner, and Doctor of Mathematical and Physical Sciences I. I. Pyatetsk-Shapiro. Formulation of the interpretational problems was greatly helped by the advice of B. Ya. Gel'chinskii. Members of the Leningrad Division of the Mathematical Institute, Academy of Sciences USSR I. Ya. Azbel' and T. B. Yanovskii, and of the Institute of Physics of the Earth, Academy of Sciences USSR V. P. Valyus and N. P. Grudev allowed us to publish a series of programs and calculations. Valuable suggestions concerning the contents of Chapter I were made by A. A. Gvozdev and Z. A. Yanson.

The authors thank the above specialists, without whose help the work could not have been completed.

We must also express our gratitude to our fellow workers in the laboratories of the Institute of Physics of the Earth, Academy of Sciences USSR, R. V. Abramova, E. G. Dirdovska, M. G. Koldaeva, and L. N. Pozharska, who performed all the technical work in the analysis of calculations and the preparation of the book.

CONTENTS

NOTATION

a — longitudinal-wave velocity

b — transverse-wave velocity

b_0 — transverse-wave velocity at the surface of a crystalline earth's crust

b_M — transverse-wave velocity in the mantle directly under the Mohorovicic discontinuity

$b(\tilde{z})$ — minimum velocity of transverse waves in the mantle

$\bar{b}$ — mean transverse-wave velocity in the crust

C_k — group velocity of the k-th mode (1.62)*

c_k^L — coefficients of the expansion of the displacement spectra of Love waves in characteristic functions (1.35)

C_{kl}^R — coefficients of the expansion of the displacements in Rayleigh waves in characteristic functions (1.42)

E — dipole parameter (3.3)

F_l — component of volume forces at the source along the l : l = T is the horizontal component, l = z is the vertical component (1.11)

$\tilde{F}_l$ — space-time spectrum (the double Fourier transform) of F_l (1.11)

G — dipole parameters (3.3)

H — thickness of the earth's crust

h — focal depth (1.56)

J_L — norming factor for the calculation of Love-wave intensity (1.36)

J_R — norming factor for the calculation of Rayleigh-wave intensity (1.43)

k — mode number (characteristic value), k = 1, 2,...

$k_L(p)$ — mode number of Love wave characteristic values of the operator (1.21) − (1.22) existing for a given p (1.34)

$k_R(\xi)$ — mode number of Rayleigh waves [characteristic values of the operator (1.16)−(1.20)] existing for given ξ (1.40)

L — proportionality coefficient in similarity laws (3.10)

M — magnitude with respect to surface waves (11.1)

m — three-dimensional frequency corresponding to the variable x (1.9)

n — three-dimensional frequency corresponding to the variable y (1.9)

p — time frequency corresponding to the variable t; the circular frequency ($p = 2\pi / T$) (1.9)

$p_{\Gamma k}$ — limiting frequency (the minimum possible) of the k-th mode

p_k^2 — k-th characteristic value of the operator (1.16-1.20)

p — group-velocity frequency extremum (Airy phase) (1.64)

R_0 — earth's radius

r — cylindrical coordinate; epicentral distance (1.11)

r_k — auxiliary function used in the calculation V_k, (2.4)

_____

*The numbers in parentheses are those of the formulas in which the notation being described is used for the first time.

T — vibration period

t — current time (1.1)

u_q — displacement component in the direction of the q axis ($q = x,\ y,\ z;\ q = \varphi,\ r,\ z$),

$q = \varphi$ — tangential component (1.1)

$q = z$ — vertical component

$q = r$ — radial component

u_{qQ} — displacement component in the direction of the q axis disturbed by the component of the force at the source in the direction of the Q axis

$Q = L$ — tangential component of the force

$Q = R$ — radial component of the force

$Q = Z$ — vertical component of the force (1.23)

U_q — time spectrum (Fourier transform) of the component u_q of the displacement in the direction of the q axis (1.49)

u_φ^k — displacement in the k-th mode of a Love wave (1.60)

U_φ^k — time spectrum of the displacement in the k-th mode of a Love wave (1.60)

$\tilde{V}_k(z)$ — k-th characteristic function of the operator (1.21)-(1.22)

$\overline{V}_k$ — frequency characteristic of the medium for the k-th mode (a concentrated force for the source)(3.1)

$\overline{V}_{dk}$ — frequency characteristic for the k-th mode (a dipole with moment for the source) (3.3)

v_k — phase velocity of the k-th mode

$v(\tau)$ — Airy function

w_k — k-th characteristic function of the operator (1.16)-(1.20)

$x,\ y$ — horizontal coordinates (1.1)

Z — depth at which the medium becomes homogeneous

z — vertical coordinate; depth of the point of observation (1.1)

$\hat{z}$ — initial point for the integration of equations (2.3) and (2.4), (2.7)

$\overline{z}_k$ — depth of penetration of the k-th mode

$\tilde{z}$ — depth of waveguide axis (layer of lowered velocity)

α — dipole parameter — azimuth of the dip of the plane of discontinuity (1.58)

γ — dipole parameter — dip of the plane of discontinuity (1.58)

δ — source parameter — azimuth of the horizontal projection of the force (1.7)

$\delta(z - h)$ — Dirac delta function (1.56)

θ_k — exhaustion function for Eq. (2.1) (2.2)

λ — Lamé parameter (1.1)

μ — Lamé parameter (1.1)

ξ — three-dimensional frequency corresponding to the variable r (1.10)

ξ_k — wave number for the k-th mode ($\xi_k = 2\pi/v_k T$)

ξ_k^2 — k-th characteristic value of the operator (1.21)-(1.22)

ρ — density (1.1)

σ_z — normal vertical stress (1.4)

$\tau_{xz},\ \tau_{yz}$ — tangential stresses (1.5)-(1.6)

τ — argument of the Airy function (1.65)

φ — cylindrical coordinate; azimuth with epicenter at the station (1.10)

$\Psi(p)$ — time spectrum of the source (1.58)

$\tilde{\omega}_k$ — k-th characteristic function of the operator (1.16)-(1.20)

CHAPTER I

THE THEORY OF LOVE WAVES

§1. Basic Formulas*

In this section the displacements in Love waves generated by a nonstationary asymmetric source in a three-dimensional medium are expressed in terms of the solution of the one-dimensional problem considered in Section 2 (i.e., in terms of the characteristic functions of a Sturm-Liouville operator).

1. Statement of the Problem

We start with the equations [47]

$$(\lambda + 2\mu)\frac{\partial^2 u_x}{\partial x^2} + \mu\frac{\partial^2 u_x}{\partial y^2} + \frac{\partial}{\partial z}\left(\mu\frac{\partial u_x}{\partial z}\right) + (\lambda + \mu)\frac{\partial^2 u_y}{\partial x \partial y} + $$
$$+ (\lambda + \mu)\frac{\partial^2 u_z}{\partial x \partial z} + \frac{\partial \mu}{\partial z}\frac{\partial u_z}{\partial x} = \rho\frac{\partial^2 u_x}{\partial t^2} - F_x; \tag{1.1}$$

$$(\lambda + \mu)\frac{\partial^2 u_x}{\partial x \partial y} + (\lambda + 2\mu)\frac{\partial^2 u_y}{\partial y^2} + \mu\frac{\partial^2 u_y}{\partial x^2} + \frac{\partial}{\partial z}\left(\mu\frac{\partial u_y}{\partial z}\right) + $$
$$+ (\lambda + \mu)\frac{\partial^2 u_z}{\partial y \partial z} + \frac{\partial \mu}{\partial z}\frac{\partial u_z}{\partial y} = \rho\frac{\partial^2 u_y}{\partial t^2} - F_y; \tag{1.2}$$

$$(\lambda + \mu)\frac{\partial}{\partial z}\left(\frac{\partial u_x}{\partial x} + \frac{\partial u_y}{\partial y}\right) + \frac{\partial \lambda}{\partial z}\left(\frac{\partial u_x}{\partial x} + \frac{\partial u_y}{\partial y}\right) + \frac{\partial}{\partial z}\left[(\lambda + 2\mu)\frac{\partial u_z}{\partial z}\right] + $$
$$+ \mu\left(\frac{\partial^2 u_z}{\partial x^2} + \frac{\partial^2 u_z}{\partial y^2}\right) = \rho\frac{\partial^2 u_z}{\partial t^2} - F_z. \tag{1.3}$$

Here x, y, and z are Cartesian coordinates; t is the time; u_x, u_y, and u_z are the components of the displacement; F_x, F_y, and F_z are the components of the body force in the direction of the x, y, and z axes respectively; λ, and μ are the Lamé parameters; ρ is the density; λ, μ, and ρ are piecewise-continuous functions of z for $0 < z \leq Z$; these functions are constant for $z > Z$, while the velocity of volume waves

$$a = \sqrt{\frac{\lambda + 2\mu}{\rho}}, \quad b = \sqrt{\frac{\mu}{\rho}}$$

has its maximum value for $z > Z$.

---

*See the footnote on p. viii.

Equations (1.1)–(1.3) are easily derived from the general equations of elastic vibrations [47] with the simplification that λ, μ and ρ depend only on z.

The boundary conditions are:

$$\sigma_z \equiv (\lambda + 2\mu)\frac{\partial u_z}{\partial z} + \lambda\left(\frac{\partial u_x}{\partial x} + \frac{\partial u_y}{\partial y}\right) = 0, \tag{1.4}$$

$$\tau_{xz} \equiv \mu\left(\frac{\partial u_x}{\partial z} + \frac{\partial u_z}{\partial x}\right) = 0, \tag{1.5}$$

$$\tau_{yz} \equiv \mu\left(\frac{\partial u_y}{\partial z} + \frac{\partial u_z}{\partial y}\right) = 0 \tag{1.6}$$

for z = 0.

In addition to this, the stresses σ_z, τ_{xz}, τ_{yz} and the displacements u_x, u_y and u_z are continuous everywhere including the point of discontinuity of λ, μ and ρ. The initial conditions are:

$$u_x = u_y = u_z = \frac{\partial u_x}{\partial t} = \frac{\partial u_y}{\partial t} = \frac{\partial u_z}{\partial t} = 0$$

for t = 0.

We wish to find the principal parts of the displacements in the region z > 0 at large distances r from the z axis. In the present work we investigate only horizontal tangential displacements (i.e., the component u_φ of the displacement in the cylindrical coordinates r, φ, z). On F_x, F_y, and F_z we impose the following limitations:

1)
$$\begin{aligned} F_x(x, y, z, t) &= F_r(r, z, t)\cos\delta, \\ F_y(x, y, z, t) &= F_r(r, z, t)\sin\delta, \end{aligned} \right\} \tag{1.7}$$

$$F_z = F_z(r, z, t), \tag{1.8}$$

where δ is a constant, and

2) in the expressions

$$F_q(x, y, z, t) = \int_{-\infty}^{+\infty}\int_{-\infty}^{+\infty}\int_{-\infty}^{+\infty} \widetilde{F}_q(\sqrt{m^2 + n^2}, p, z)e^{i\,(mx+ny+pt)}\,dm\,dn\,dp, \tag{1.9}$$

$$(q = x;\ y;\ z)$$

the Fourier transform is bounded and decreases exponentially when m, n, and p increase, starting from some constant values.

Introducing the new variables

$$\begin{aligned} m &= \xi\cos\alpha, \\ n &= \xi\sin\alpha \end{aligned} \right\} \tag{1.10}$$

and the coordinates

$$x = r \cos \varphi,$$
$$y = r \sin \varphi,$$

in (1.9) and using (1.7) and (1.8), we obtain

$$F_l = 2\pi \int\limits_{-\infty}^{+\infty} \int\limits_{0}^{+\infty} \widetilde{F}_l (\xi, p, z) J_0 (\xi r) e^{ipt} \xi \, d\xi \, dp,$$

$$(l = T, z).$$

2. Expressions for the Displacements in Terms of the Solution of Two Plane Problems

We reduce the solution of the problem stated above to that of the solution of two independent plane problems, in one of which the forces and displacements are perpendicular, and in the other parallel, to a certain plane containing the z axis. To do this we express the displacements u_x, u_y, and u_z in the form

$$u_q (x, y, z, t) = \int\limits_{-\infty}^{+\infty} \int\limits_{-\infty}^{+\infty} \int\limits_{-\infty}^{+\infty} \widetilde{u}_q (m, n, p, z) e^{i(mx+ny+pt)} \, dm \, dn \, dp, \qquad (1.12)$$

$$(q = x, y, z),$$

and then introduce the transformations

$$\widetilde{u}_x = \frac{1}{\xi} [\omega_T \cos(\alpha - \delta) + \omega_z] m + \frac{1}{\xi} V \sin(\alpha - \delta) n, \qquad (1.13)$$

$$\widetilde{u}_y = \frac{1}{\xi} [\omega_T \cos(\alpha - \delta) + \omega_z] n - \frac{1}{\xi} V \sin(\alpha - \delta) m, \qquad (1.14)$$

$$\widetilde{u}_z = - i [\omega_T \cos(\alpha - \delta) + w_z]. \qquad (1.15)$$

These transformations are selected so that the vector with components ω_l, w_l, ($l = T$, z) and the function V satisfy two mutually independent systems of equations and boundary conditions. We assume, for the moment arbitrarily, that (1.12) can be twice differentiated with respect to x, y, z, and t under the sign of integration; we substitute (1.9) and (1.12) in (1.1)–(1.6); we apply the transformations (1.10) and (1.13)–(1.15) to the resulting equations for the $\widetilde{u}_q$. Then after some relatively simple calculations we obtain the following two operators.

1) An operator containing ω_l, w_l ($l = T$, z);

$$\frac{d}{dz} \left[\mu \frac{d\omega_l}{dz} + \xi\mu w_l \right] + \xi\lambda \frac{dw_l}{dz} + \omega_l [p^2\rho - \xi^2 (\lambda + 2\mu)] = - N_l, \qquad (1.16)$$

$$\frac{d}{dz} \left[(\lambda + 2\mu) \frac{dw_l}{dz} - \xi\lambda\omega_l \right] - \xi\mu \frac{d\omega_l}{dz} + w_l [p^2\rho - \xi^2\mu] = - M_l, \qquad (1.17)$$

$$\left. \begin{array}{ll} N_T = \widetilde{F}_T, & M_T = 0, \\ N_z = 0, & M_z = i\widetilde{F}_z. \end{array} \right\} \qquad (1.18)$$

For z = 0

$$(\lambda + 2\mu)\frac{dw_l}{dz} - \xi\lambda\omega_l = 0, \tag{1.19}$$

$$\mu\left(\frac{d\omega_l}{dz} + \xi w_l\right) = 0. \tag{1.20}$$

When $z \to \infty$ we have $w_l \to 0$ and $\omega_l \to 0$. The left-hand sides of the last two equations and w_l and ω_l are continuous.

2) An operator containing V:

$$\frac{d}{dz}\left(\mu\frac{dV}{dz}\right) + V(\rho p^2 - \xi^2\mu) = -\widetilde{F}_T. \tag{1.21}$$

For z = 0 we have

$$\mu\frac{dV}{dz} = 0. \tag{1.22}$$

When $z \to \infty$ $V \to 0$.

The functions $z, \mu dV/dz$, and V are continuous for all z.

These two operators determine the solution of the two plane problem given above.

Replacing λ, μ, and ρ by step functions, we obtain the operators for a system of homogeneous layers, usually considered in surface-wave theory; the first corresponds to Rayleigh waves, the second to Love waves.

We substitute (1.13)-(1.15) in (1.12), make the change of variable (1.10), integrate with respect to α, and use the fact that $\widetilde{F}_T$, and $\widetilde{F}_Z$, and so w_l, ω_l and V are independent of α. Then from known relations we obtain

$$\int_0^{2\pi} \exp[i\xi r \cos\alpha]\, d\alpha = 2\pi J_0(\xi r),$$

$$\int_0^{2\pi} \exp[i\xi r \cos\alpha]\, \begin{matrix}\sin\alpha \\ \cos\alpha\end{matrix}\, d\alpha = \begin{matrix}0 \\ 2\pi i\end{matrix} J_1(\xi r),$$

$$\int_0^{2\pi} \exp[i\xi r \cos\alpha]\, \begin{matrix}\sin \\ \cos\end{matrix}\, (\delta - \alpha)\cos\alpha\, d\alpha = 2\pi\, \begin{matrix}\sin\delta \\ \cos\delta\end{matrix}\left[J_0(\xi r) - \frac{1}{\xi r}J_1(\xi r)\right],$$

$$\int_0^{2\pi} \exp[i\xi r \cos\alpha]\, \begin{matrix}\sin \\ \cos\end{matrix}\, (\alpha - \delta)\sin\alpha\, d\alpha = 2\pi\, \begin{matrix}\cos\delta \\ \sin\delta\end{matrix}\, J_1(\xi r)\frac{1}{\xi r}.$$

After referring the displacements u_r, u_φ, and u_z to cylindrical coordinates and carrying out a series of calculations we now have

$$u_q = u_{qT} + u_{qZ}$$

$$(q = r,\, \varphi,\, z)$$

$$u_q = \int_{-\infty}^{+\infty} e^{ipt} U_{qQ}\, dp\,. \tag{1.23}$$

Q indicates a force component and q is a displacement component in the source.

Here

$$U_{rT} = 2\pi \cos{(\delta - \varphi)} \left\{ \int_0^\infty \omega_T(\xi, p, z) [J_0(\xi r) - (1/\xi r) J_1(\xi r)] \xi d\xi + \frac{1}{r} \int_0^\infty V(\xi, p, z) J_1(\xi r) d\xi \right\} \quad (1.24)$$

$$U_{\varphi T} = 2\pi \sin{(\delta - \varphi)} \left\{ \int_0^\infty V(\xi, p, z)[J_0(\xi r) - (1/\xi r) J_1(\xi r)] \xi d\xi + \frac{1}{r} \int_0^\infty \omega_T(\xi, p, z) J_1(\xi r) d\xi \right\} \quad (1.25)$$

$$U_{zT} = 2\pi \cos{(\delta - \varphi)} \int_0^\infty w_T(\xi, p, z) J_1(\xi r) \xi d\xi \quad (1.26)$$

$$U_{rZ} = 2\pi i \int \omega_z(\xi, p, z) J_1(\xi r) \xi d\xi \quad (1.27)$$

$$U_{\varphi Z} = 0 \quad (1.28)$$

$$U_{zZ} = -2\pi i \int_0^\infty \omega_z(\xi, p, z) J_0(\xi r) \xi d\xi . \quad (1.29)$$

We see that the displacements u_{qz} due to the vertical force F_z, in our case, i.e., when the source and the medium are axially symmetric, are polarized in the rz plane. They represent longitudinal, transverse SV, and Rayleigh waves, the properties of which are determined by the operator (1.16)-(1.20).

The displacements U_{qz}, due to the horizontal force F_T, are polarized in a complex way. At large distances, r, however, the vibrations in the rz plane are generated independently of vibrations in directions perpendicular to this plane. The first (with components u_{rT} and u_{zT}) contain the same waves as those in the case of a vertical force and are controlled by the same operator. At any given point, u_{rT} and u_{zT} are proportional to the radial component F_R (parallel to the rz plane) of the force F_T and are given by the relations

$$U_{rT} \approx U_{rR} = 2\pi \cos{(\delta - \varphi)} \int_0^\infty \omega_T(\xi, p, z) [J_0(\xi r) - (1/\xi r) J_1(\xi r)] \xi d\xi . \quad (1.30)$$

$$U_{zT} \approx U_{zR} = 2\pi \cos{(\delta - \varphi)} \int_0^\infty w_T(\xi, p, z) J_1(\xi r) \xi d\xi . \quad (1.31)$$

The component $u_{\varphi T}$ represents the SH body wave and the Love wave. At any given point $U_{\varphi T}$ is proportional to the tangential projection F_L (perpendicular to the rz plane) of the force F_T. We have the relation

$$U_{\varphi T} \approx U_{\varphi L} = 2\pi \sin{(\delta - \varphi)} \int_0^\infty V(\xi, p, z) [J_0(\xi r) - (1/\xi r) J_1(\xi r)] \xi d\xi . \quad (1.32)$$

3. Expressions for Displacements in Terms of Characteristic Functions of the Corresponding Operators

The characteristic functions of the problem (1.21) and (1.22) are the nonzero solutions of this problem for $\widetilde{F}_T = 0$ [11, 23]. Such solutions exist only for certain values of $\xi^2(p^2)$ called characteristic values of the problem.

For fixed p^2, there is a finite number of characteristic values $\xi^2 = \xi^2_{kL}(p^2)$ (a discrete spectrum) in the interval*

$$\frac{p^2}{b^2_{\min}} > \xi^2_{kL} \geqslant \frac{p^2}{b^2(Z+0)}, \tag{1.33}$$

while the interval $-\infty \leq \xi^2_L \leq p^2/[b^2(Z+0)]$ is a continuous spectrum of characteristic values. The characteristic functions for the continuous and discrete spectrum will be denoted by $\widetilde{V}(\xi^2_L, z)$ and $\widetilde{V}_k(\xi^2_{kL}, z)$ respectively. From the general theory of characteristic functions, the solution of (1.21) for $\widetilde{F}_T \neq 0$ can be expressed in the form

$$V(\xi, z) = \sum_{k=1}^{k=k_L(p)} c^L_k \widetilde{V}_k(\xi^2_{kL}, z) + \int_{-\infty}^{p^2/b^2(Z+0)} c^L(\nu) \widetilde{V}(\nu, z) d\nu, \tag{1.34}$$

where

$$c^L_k = \frac{J_L}{\xi^2 - \xi^2_{kL}} \int_0^\infty \widetilde{F}_T(\xi, p, z) \widetilde{V}_k(\xi^2_{kL}, z) dz, \tag{1.35}$$

$$J_L = \left[\int_0^\infty \mu |\widetilde{V}_k|^2 dz \right]^{-1}, \tag{1.36}$$

$$c^L(\nu) = \frac{1}{\xi^2 - \nu} \int_0^\infty \widetilde{F}_T(\xi, p, z) \widetilde{V}(\nu, z) dz, \tag{1.37}$$

where $k_L(p^2)$ is the number of characteristic functions for a fixed value of p^2. In the derivation of these formulas we use the orthogonality conditions

$$\int_0^\infty \mu \widetilde{V}_i \widetilde{V}_j dz = 0 \quad \text{for} \quad i \neq j, \tag{1.38}$$

$$\int_0^\infty \mu \widetilde{V}(\xi^2, z) \widetilde{V}(\nu, z) dz = \delta(\xi^2 - \nu), \tag{1.39}$$

where δ is the Dirac delta function.

We set $N_l = 0$ in (1.16) and $M_l = 0$ in (1.17) and fix the values of ξ^2. Then the characteristic values of the problem, i.e., the values of $p^2(\xi^2)$ for which (1.16)-(1.20) have nonzero solutions, form a continuous spectrum in the interval

$$\xi^2 b^2(Z+0) < p^2_R < \infty$$

---

*In the following sections, ξ_{kL} will be denoted simply by ξ_k.

and a discrete spectrum in the interval

$$\xi^2 b^2 (Z + 0) \geqslant p_{kR}^2 (\xi^2) > \xi^2 v_R^2,$$

where v_R is the minimum velocity of boundary waves corresponding to a homogeneous half-space with constants equal to $a(z)$, $b(z)$, and $\rho(z)$ at some depth. The lower limit is obtained here from physical considerations; it is not necessary to prove that it is negative.

The corresponding nonzero solutions of (1.16)–(1.20) (the characteristic functions) will be denoted by $(\widetilde{\omega}, \widetilde{w})$ and $(\widetilde{\omega}_k, \widetilde{w}_k)$.

For nonzero N_l or M_l, the solution of (1.16)–(1.20) can be written in terms of the characteristic functions as follows:

$$\omega_l = \sum_{k=0}^{k=k_R(\xi)} c_{kl}^R \widetilde{\omega}_k + \int_{\xi^2 b^2 (Z+0)}^{\infty} c_l^R (\nu) \widetilde{\omega}(\nu, z) \, d\nu, \tag{1.40}$$

$$w_l = \sum_{k=0}^{k=k_R(\xi)} c_{kl}^R \widetilde{w}_k + \int_{\xi^2 b^2 (Z+0)}^{\infty} c_l^R (\nu) \widetilde{w}(\nu, z) \, d\nu. \tag{1.41}$$

Here

$$c_{kl}^R = \frac{-J_R}{p^2 - p_{kR}^2} \int_0^{\infty} [N_l (\xi, p, z) \widetilde{\omega}_k + M_l (\xi, p, z) \widetilde{w}_k] \, dz, \tag{1.42}$$

$$J_R = \left[\int_0^{\infty} \rho (|\widetilde{\omega}_k|^2 + |\widetilde{w}_k|^2) \, dz \right]^{-1}, \tag{1.43}$$

$$c_l^R = \frac{-1}{p^2 - \nu} \int_0^{\infty} [N_l (\xi, p, z) \widetilde{\omega}(\nu, z) + M_l (\xi, p, z) \widetilde{w}(\nu, z)] \, dz \tag{1.44}$$

and $k_R(\xi)$ is the number of characteristic values for a given ξ.

In the following, we will consider another value of ξ_{kR} — that value of ξ for which $p = p_{kR}$; we denote this value by $k_R(p)$ — it is the number of ξ_{kR}^2 for which $p_{kR} = p$. In the derivation of these formulas we use the orthogonality conditions

$$\int_0^{\infty} \rho [\widetilde{\omega}_i \widetilde{\omega}_j + \widetilde{w}_i \widetilde{w}_j] \, dz = 0 \quad \text{for} \quad i \neq j, \tag{1.45}$$

$$\int_0^{\infty} \rho [\widetilde{\omega}(\nu) \widetilde{\omega}(p^2) + \widetilde{w}(\nu) \widetilde{w}(p^2)] \, dz = \delta (p^2 - \nu). \tag{1.46}$$

4. An Asymptotic Representation of the Displacement Spectrum

Asymptotic estimates of integrals differing from (1.24)–(1.32) only in the form of ω_l, w_l, and V were obtained in [4, 9, 15-18, 27]; in these articles expressions were used for ω_l, w_l, and V corresponding to piecewise-constant functions λ, μ, and ρ (i.e., a layered-homogeneous medium).

It is easily seen that, for piecewise-continuous λ, μ, and ρ, the functions ω_l, w_l, and V retain the same properties that were used in the derivation of the asymptotic estimates in [9, 17].

These properties are the following:

a) When the above limitations are imposed on $\tilde{f}'_q$, the coefficients c_k^L, c_{kl}^R, and thus ω_l, w_l, and V, decrease exponentially when the corresponding parameter increases beginning with some sufficiently large value of this parameter;

b) $k_L(p)$ and $k_R(p)$ are bounded for all finite values of p, although they increase with increasing p;

c) V and ω_l are odd and w_l is an even function of ξ; hence the coefficients of $J_1(\xi r)$ under the integration sign in (1.24)–(1.32) are even and those of $J_0(\xi r)$ are odd functions of ξ;

d) The only singularities of C^L, C^R are the branch points $\xi = \pm p/b(Z+0)$, while the only singularities of C^R are these same branch points and also the branch points $\xi = \pm p/a(Z+0)$; at infinity these functions behave in the same way as C_k^L and C_k^R [23].

The properties enumerated above are sufficient to enable us to extend the results obtained in [9, 16, 17] to the functions ω_l, w_l, and V.

The proofs in [9, 17, 18] of the correctness of the solution [in particular the validity of the differentiation of integrals of the type (1.12) under the sign of integration] can be extended to apply to the solution (1.23) without change.

For stationary oscillations U_{qQ} described by (1.24)–(1.32) we can use the following asymptotic formulas:

$$\int_0^\infty \frac{f(\xi)}{\xi^2 - \beta^2} \, J_\nu(\xi r) \xi \, d\xi = f(\beta) \sqrt{\frac{\pi}{2\beta r}} \, \exp\left[-i\beta r - i(-1)^\nu \frac{\pi}{4} \right] + 0\left(r^{-3/2} \right) \tag{1.47}$$

$$\int_0^\infty \frac{f(\xi)}{p_{kR}^2(\xi) - p^2} \, J_\nu(\xi r) \xi \, d\xi = f(\beta) \sqrt{\frac{\pi}{2\beta r}} \, \frac{d\beta}{dp} \, \exp\left[-i\beta r - i(-1)^\nu \frac{\pi}{4} \right] + 0\left(r^{-3/2} \right). \tag{1.48}$$

Here $\nu = 0$ or 1, β is equal to ξ_{kL} or ξ_{kR}; $f(\beta)$ is the corresponding characteristic function. The first estimate is for even and the second for odd functions $f(\xi)$.*

According to the estimates in these references, a disturbance corresponding to a continuous spectrum [the integral terms in (1.34), (1.40), (1.41)] decreases in strength at infinity as $r^{-3/2}$.

Substituting (1.37), (1.43), and (1.44) in (1.24)–(1.32), using (1.47) and (1.48), and writing U_q for the spectrum of u_q ($U_q = U_{qR} + U_{qZ} + U_{qL}$), we finally obtain the following results.

Rayleigh Waves

$$U_r(r, \varphi, z, p) = \sum_{k=1}^{k_R(p)} \frac{D_{kR}}{\sqrt{\xi_{kR} r}} \, \widetilde{\omega}_k(\xi_{kR}, p, z) \exp(-i\xi_{kR} r) + 0(r^{-3/2}), \tag{1.49}$$

---

*The integrals on the left in (1.47) and (1.48) are to be interpreted as Cauchy principal values. In the derivation of these estimates in [9, 18], a system of characteristic functions is used on the right-hand side that guarantees that the conditions at infinity will be satisfied. Hence the estimates are valid only for the indicated choice of β and $f(\beta)$. The references cited give detailed derivations.

$$U_z(r, \varphi, z, p) = \sum_{k=1}^{k_R(p)} \frac{iD_{kR}}{\sqrt{\xi_{kR}r}} \widetilde{w}_k(\xi_{kR}, p, z) \exp(-i\xi_{kR}r) + 0(r^{-3/2}), \tag{1.50}$$

$$D_{kR} = \frac{\xi_{kR}}{p} \frac{d\xi_{kR}}{dp} e^{i\frac{\pi}{4}} \sqrt{2\pi^3} \int_0^\infty [i\widetilde{F}_T \cos(\delta - \varphi)\widetilde{\omega}_k + \widetilde{F}_z\widetilde{w}_k] dz \tag{1.51}$$

where $\widetilde{\omega}_k$, and $\widetilde{w}_k$ are normed so that

$$J_R = 1. \tag{1.52}$$

Love Waves

$$U_\varphi(r, \varphi, z, p) = \sum_{k=1}^{k_L(p)} \frac{D_{kL}}{\sqrt{\xi_{kL}r}} \widetilde{V}_k(\xi_{kL}, p, z) \exp[-i\xi_{kL}r] + 0(r^{-3/2}), \tag{1.53}$$

$$D_{kL} = e^{-i\frac{\pi}{4}} \sqrt{2\pi^3} \int_0^\infty \widetilde{F}_T(\xi_{kL}, p, z) \sin(\delta - \varphi)\widetilde{V}_k(\xi_{kL}, p, z) dz, \tag{1.54}$$

with $\widetilde{V}_k$ normed so that

$$J_L = 1. \tag{1.55}$$

Here disturbances corresponding to the continuous spectrum are in the remainder term.

The last three formulas also determine the relation between the spectrum of the disturbances in Love waves and the functions ξ_{kL} and $\widetilde{V}_k$, which give the solution of the problem considered in Section 2.

In what follows we also need to know the form of (1.53) for two specific sources.

1) Forces distributed on the plane z = h

$$\widetilde{F}_T = \delta(z - h)\widetilde{F}^*(p, \xi). \tag{1.56}$$

Here $\delta(z - h)$ is the Dirac delta function and $\widetilde{F}^*(p, \xi)$ is given by (1.11) as before.

Substituting (1.56) and (1.54) in (1.53) we have

$$U_\varphi = e^{-i\frac{\pi}{4}} \sqrt{2\pi^3} \sum_{k=1}^{k_L(p)} \widetilde{F}(p, \xi_{kL}) \sin(\delta - \varphi) \frac{\widetilde{V}_k(\xi_{kL}, p, z)\widetilde{V}_k(\xi_{kL}, p, h)}{\sqrt{\xi_{kL}r}} \times \exp(-i\xi_{kL}r). \tag{1.57}$$

2) A dipole with a moment applied at the point r = 0, z = h* [31]

Displacements generated by a dipole are obtained by differentiating displacements generated by a simple force with respect to the distance in some direction $\overline{l}$ depending on the orientation of the dipole. In the

---

*The estimates used here hold for point sources only when the spectrum is sufficiently rapidly damped when p increases. Instead of this we can assume that the source is actually not perfectly concentrated, but that it is concentrated for the periods under consideration. These questions are considered in detail in [9, 18].

case of a dipole with a moment, $\overline{l}$ is perpendicular to the direction of the force and in the direction of the normal to the plane of discontinuity.

We denote by γ and α the inclination to the vertical and the azimuth of the direction of $\overline{l}$ (the angle and the azimuth of the dip of the plane of discontinuity), by φ the azimuth of the point of observation, and by $\psi(p)$ the time spectrum of the source. Then it is easily seen that

$$U\varphi = e^{-i\pi/4} \sqrt{2\pi^3} \sum_{k=1}^{k_L(P)} \frac{\Psi(P)}{\sqrt{\xi_{kL} r}} \sin(\delta - \varphi) \tilde{V}_k(z) \left[-i \sin\gamma \cos(a-\varphi) \tilde{V}_k(h) \xi_{kL} - \cos\gamma \tilde{V}'_k(h)\right] \exp[-i\xi_{kL} r]. \quad (1.58)$$

Here

$$\tilde{V}'_k(h) = \frac{d}{dz} \tilde{V}_k(z) \quad (z = h). \quad (1.59)$$

5. Modes of Surface Waves and their Phase Velocities

In our notation, displacements due to stationary vibrations with frequency p are equal to the corresponding spectrum form (1.49), (1.50), (1.54) multiplied by e^{ipt}. Hence Rayleigh waves are described by the sum of terms with factors $\exp\{i(pt - \xi_{kR}r)\}$ while Love waves are described by the sum of terms with the factor $\exp\{i(pt - \xi_{kL}r)\}$.

These terms are called the modes of a surface wave and k is the number of the mode. It is easily seen that the phase velocity of the k-th mode of Rayleigh and Love waves are, respectively,

$$v_{kR} = \frac{p}{\xi_{kR}}, \quad v_{kL} = \frac{p}{\xi_{kL}}.$$

The dependence ξ_{kR} and ξ_{kL} on the frequency is due to dispersion of phase velocities. The graph of the dependence of v_{kR}, and v_{kL} on the frequency or on the period is called a branch of the phase-velocity dispersion curve.

6. An Asymptotic Formula for Nonstationary Displacements

For completeness we give the known formulas expressing nonstationary displacements $u_\varphi^{(k)}$ in the k-th mode of surface waves for larger r in terms of the spectra of these displacements:

$$u_\varphi^{(k)} = 2 \int_0^\infty U_\varphi^{(k)} e^{ipt} dp. \quad (1.60)$$

It is known that, far from the extrema of the group velocity $dp/d\xi_k$, i.e., where $|d^2\xi_k/dp^2|$ is not very small, we can use approximations to the integrals (1.60) obtained by the stationary-phase method [16 - 18]:

$$\int_0^\infty \Phi(p) e^{i(pt - \xi_k r)} dp = \sum_j \Phi(p_j) \sqrt{\frac{2\pi}{r \left|\frac{d^2\xi_k}{dp^2}\right|_j}} \exp\left[i(p_j t - \xi_k^j r) \pm i\frac{\pi}{4}\right] + 0(r^{-3/2}). \quad (1.61)$$

The sign of $i(\pi/4)$ in the exponent is plus when $d^2\xi_k/dp^2 < 0$ and minus when $d^2\xi_k/dp^2 > 0$); the p_j are roots of the equation

$$C_k \equiv \frac{dp}{d\xi_k} = \frac{r}{t} . \tag{1.62}$$

For a given value of r these roots are determined in practice as follows: We take a value of t and find the abscissas p_j of the points on the graph of the group velocity $C_k(p)$ where this velocity is equal to r/t.

Using (1.61) and substituting (1.53) and (1.54) into (1.60), we obtain the final expression for the nonstationary displacements in the k-th mode of Love waves far from the group-velocity extrema:

$$u_\varphi^{(k)} = \frac{4\pi^2}{r} e^{\pm i\frac{\pi}{4} - i\frac{\pi}{4}} \sum_j \frac{1}{\sqrt{\xi_{kL}^j}} \left[\int_0^\infty \widetilde{F}_T^j \sin(\delta - \varphi) \widetilde{V}_k^j dz \right] \widetilde{V}_k^j \frac{\exp[i(p_j t - \xi_{kL}^j r)]}{\sqrt{\left| \dfrac{d^2\xi_{kL}(p_j)}{dp^2} \right|}} + O(r^{-3/2}). \tag{1.63}$$

Here ξ_{kL}^j, $\widetilde{F}_T^j$, $\widetilde{V}_k^j$, $|d^2\xi_{kL}(p_j)/dp^2|$ correspond to the values of p_j that are roots of (1.62). The sign to be used in the exponent is specified above.

To find $u_\varphi^{(k)}$ at the extremal points $\bar{p}$ where $d^2\xi_k/dp^2 = 0$ (but $d^3\xi_k/dp^3 \neq 0$), we use the representation (1.60) and employ the Airy function [4, 16]:

$$\int_0^\infty \Phi(p) e^{i(pt - \xi_k r)} dp = \frac{2\sqrt{\pi}\, \Phi(\bar{p}) e^{i(\bar{p}t - \bar{\xi}_k r)}}{\sqrt[3]{\dfrac{r}{2} \left| \dfrac{d^3\xi_k(\bar{p})}{dp^3} \right|}} v(\tau), \tag{1.64}$$

where v is the Airy integral with the argument

$$\tau = \frac{t - r \left| \dfrac{d\xi_k(\bar{p})}{dp} \right|}{\sqrt[3]{\dfrac{r}{2} \left| \dfrac{d^3\xi_k(\bar{p})}{dp^3} \right|}} . \tag{1.65}$$

Now using (1.64) and (1.65), we obtain for the nonstationary displacement $\bar{u}_\varphi^{(k)}$ in the k-th mode of a Love wave with extremal point $\bar{p}$ the formula

$$\bar{u}_\varphi^{(k)} = \frac{8\pi^2}{\sqrt{2}} \frac{1}{r^{5/6}} e^{-i\frac{\pi}{4}} \frac{\left[\int_0^\infty \overline{\widetilde{F}}_T \sin(\delta - \varphi) \overline{\widetilde{V}}_k dz \right] \overline{\widetilde{V}}_k e^{i(\bar{p}t - \bar{\xi}_{kL} r)} v(\tau)}{\sqrt{\bar{\xi}_{kL}} \sqrt[3]{\left| \dfrac{d^3\bar{\xi}_{kL}(\bar{p})}{dp^3} \right|}} . \tag{1.66}$$

Here $\bar{\xi}_{kL}$, $\overline{\widetilde{F}}_T$, $\overline{\widetilde{V}}_k$, $|d^3\xi_{kL}(\bar{p})/dp^3|$ are calculated for $p = \bar{p}$, which corresponds to an extremal point of $dp/d\xi_{kL}$; $v(\tau)$ is given by (1.65).

The accuracy of (1.63) decreases in the neighborhood of extremal points, and more accurate estimates can be found in [30]. [The function $u_\varphi^{(k)}(t)$ is finite and real, and the symbol Re in (1.60), (1.63), and (1.66) indicates that the real part of the following expression is to be taken.]

§2. Methods of Calculating the Parameters of Love Waves

1. Statement of the Problem

As was shown in Section 1 [formulas (1.21), (1.22), and (1.53)], the displacements in Love waves can be simply expressed in terms of the characteristic values and characteristic functions of the following problem.

Consider the equation

$$\frac{d}{dz}\left(\mu(z)\frac{d\widetilde{V}_k}{dz}\right) + p^2\rho(z)\widetilde{V}_k = \xi^2\mu(z)\widetilde{V}_k \tag{2.1}$$

in the interval $[0, +\infty]$, with the boundary conditions

$$\frac{d\widetilde{V}_k}{dz}\bigg|_{z=0} = 0, \quad \int_0^\infty \mu(z)\widetilde{V}_k^2\,dz < +\infty. \tag{2.1'}$$

Here p, and ξ are real parameters; $\rho(z)$ is a positive function and $\rho(z) \equiv \rho(Z)$ for $z \geq Z$; $\mu(z) = b^2(z)\rho(z)$, $b^2(z) \equiv b^2(Z+0)$ for $z \geq Z$ and $b^2(z) \leq b^2(Z+0)$ for $0 \leq z \leq Z$; $\mu\, d\widetilde{V}_k/dz$ is continuous for all z.

The values $\xi_k^2(p)$ (k = 1, 2 ...) of the parameter ξ^2 for which there are nonzero solutions $\widetilde{V}_k(z)$ of Eq. (2.1) satisfying (2.1') are called the characteristic values of the problem (2.1) and (2.1'). The functions $\widetilde{V}_k(z)$ are called characteristic functions of this problem.

It is known [11, 12, 23] that for a fixed value of p there is a finite number of characteristic values of this problem:

$$\xi_1^2(p) > \xi_2^2(p) > \cdots > \xi_k^2(p) > \cdots > \xi_{k_L}^2(p).$$

For sufficiently small values of p, there is only one characteristic value $\xi_1^2(p)$. As p increases the number of characteristic values can only increase.

It is also known that the k-th characteristic function $\widetilde{V}_k(z)$ (k = 1, 2...) has $k-1$ zeros in the interval $[0, Z]$ and decreases exponentially in the interval $[Z, +\infty]$.

In this section we describe a method of calculating all the characteristic values $\xi_k^2(p)$ and the corresponding characteristic functions $\widetilde{V}_k(z, p)$ with any initially given accuracy for a given interval of values of p.

2. The Solution of the Problem for Fixed Frequencies

In the solution of the boundary-value problem (2.1), (2.1') it is best to use the exhaustion method proposed by I. M. Gel'fand and O. V. Lokutsievskii [7]. We will use a variant of this method proposed by A. A. Abramov [1].

With any solution $\widetilde{V}(z)$ of Eq. (2.1) we associate the continuous function $\theta(z)$

$$\theta(z) = \arctan\left(\frac{\mu(z)\dfrac{d\widetilde{V}}{dz}}{\widetilde{V}}\right)^*. \tag{2.2}$$

---

*More precisely $\theta = \arctan(\mu(z)(d\widetilde{V}/dz)/N\widetilde{V})$; where N is a norming factor with dimension $(\mu)/(z)$; in the following we will assume that N = 1 and use the dimensionless quantitites, p, ρ, μ, z.

Differentiating (2.2) with respect to z and substituting the resulting expression for d/dz(μ d$\widetilde{V}$/dz) in (2.1), we obtain the equation

$$\frac{d\theta}{dz} = -\frac{1}{\mu(z)}\sin^2\theta - (p^2\rho(z) - \xi^2\mu(z))\cos^2\theta \qquad (2.3)$$

for the function $\theta(z)$ in the interval [0, + ∞].

We now write the boundary conditions (2.1') for $\theta(z)$. From (2.2) with z = 0 we obtain tan $\theta(0) = 0$. We assume that $\theta(0) = 0$. We write the second boundary condition for z = Z. For $z \geq Z$, the coefficients of Eq. (2.1) are constant, and the square-integrable solution of (2.1) is

$$\widetilde{V}(z) = B\exp\left(-\sqrt{\xi^2 - \frac{p^2}{b^2(z)}}z\right).$$

$$\theta(z) = -\text{arc tg}\left(\mu(Z)\sqrt{\xi^2 - \frac{p^2}{b^2(z)}}\right) - (k-1)\pi$$

where B is a constant.

Let ξ_k^2 be the k-th characteristic value and $\widetilde{V}_k(z)$ the k-th characteristic function of the problem (2.1), (2.1'); let $\theta_k(z)$ be the solution of Eq. (2.3) corresponding to $\widetilde{V}_k(z)$ and satisfying the condition $\theta_k(0) = 0$ and the condition (2.3') for z = Z.

We show that the number k in (2.3') coincides with the number of characteristic values ξ_k^2. At some point s let $\widetilde{V}_k(s) = 0$. It follows from (2.2) that $\theta_k(s) = -\pi/2(2n + 1)$, where n is an integer. We easily see that $\theta_k(z)$ decreases monotonically when z increases in the vicinity of s. In fact Eq. (2.3) yields $\theta_k'(s) = -1/\mu(s) < 0$. This means that the number of zeros of the characteristic function $V_k(z)$ in [0, Z] is the same as the number of points of the form $-\pi/2(2n + 1)$ in the interval $[-(k-1)\pi, 0]$. The number of such points is clearly $(k-1)$.

The converse is also easily proved, namely, that if for some value of ξ^2, the function $\theta(z)$ satisfies Eq. (2.3) and the boundary conditions for z = 0 and z = Z for some k, then this value of ξ^2 is the k-th characteristic value of the problem (2.1), (2.1').

The characteristic values ξ_k^2 are calculated by a "fitting" method: we integrate Eq. (2.3) from the right with the initial condition (2.3') and calculate $\theta(0)$. We choose the value of ξ^2 for which $\theta(0) = 0$. This value is ξ_k^2.

The method used for integrating Eq. (2.3) is described below in Paragraph 3, the fitting method is described in Paragraph 4 of this section.

To find the characteristic function $\widetilde{V}_k(z)$ we introduce a new unknown function $r_k(z)$ by the relations (see [12])

$$\mu\frac{d\widetilde{V}_k(z)}{dz} = r_k(z)\sin\theta_k(z), \quad \widetilde{V}_k(z) = r_k(z)\cos\theta_k(z).$$

We obtain the equation

$$\frac{dr_k}{dz} = -\frac{1}{2}\sin 2\theta_k\left(p^2\rho(z) - \xi^2\mu(z) - \frac{1}{\mu(z)}\right)r_k. \qquad (2.4)$$

After finding the characteristic value ξ_k^2 we solve Eq. (2.4) simultaneously with Eq. (2.3) and determine the functions $r_k(z)$, $\widetilde{V}_k(z)$ and $\widetilde{V}_k'(z)$. We calculate $\int_0^\infty \rho\widetilde{V}_k^2 dz$, $\int_0^\infty \mu\widetilde{V}_k^2 dz$, $\int_0^\infty \mu\widetilde{V}_k'^2 dz$ at the same time.

We now calculate the characteristic value ξ_k^2 more accurately by using the formula

$$\xi_k^2 = \frac{p^2 \int_0^\infty \rho \widetilde{V}_k^2 dz - \int_0^\infty \mu \widetilde{V}_k'^2 dz}{\int_0^\infty \mu \widetilde{V}_k^2 dz} \tag{2.5}$$

from the calculus of variations.

To obtain this formula we multiply Eq. (2.1) by $\widetilde{V}_K(z)$, integrate both sides of the resulting equation with respect to z from 0 to ∞, and integrate $\int_0^\infty (\mu \widetilde{V}_k')' \widetilde{V}_k dz$ by parts.

We now calculate the phase velocity $v_k = p/\xi_k$ and the group velocity $C_k = dp/d\xi_k$. For the latter calculation we use the formula

$$\frac{dp}{d\xi} = \frac{1}{p/\xi_k} \frac{\int_0^\infty \mu \widetilde{V}_k^2 dz}{\int_0^\infty \rho \widetilde{V}_k^2 dz} \tag{2.6}$$

from perturbation theory.

To obtain this formula we differentiate (2.1) with respect to the parameter p, multiply both sides by $\widetilde{V}_K(z)$, and integrate with respect to z from 0 to ∞.

We also calculate the product of the functions

$$\frac{\widetilde{V}_k(z) \, |\widetilde{V}_k(0)|}{2\xi_k \int_0^\infty \mu \widetilde{V}_k^2 dz} \quad \text{and} \quad \frac{\widetilde{V}_k'(z) \, |\widetilde{V}_k(0)|}{2\xi_k \int_0^\infty \mu \widetilde{V}_k^2 dz}$$

normed so that (1.55) holds.*

3. A Method of Solving the Initial-Value Problem for Eq. (2.3)

We integrate Eq. (2.3) from right to left, i.e., from z = Z to zero. By the same token we calculate the function $\theta(z)$ corresponding to an exponentially decreasing solution of Eq. (2.1) for any ξ.

When the deviation ξ from the characteristic value is small this solution differs only slightly from the characteristic function. If we integrate Eq. (2.3) from left to right with the initial condition $\theta(0) = 0$, then when the deviation ξ from a characteristic value is small the resulting function $\theta(z)$ corresponds to an exponentially increasing solution of Eq. (2.1), and this solution can differ greatly from a characteristic function close to the point z = Z.

For many values of the parameter ξ and p there is no reason to integrate Eq. (2.3) and (2.4) over the whole interval from z = Z to 0, since the value of the characteristic function $\widetilde{V}_K(Z)$ can be very small compared

---

*The norm factor $J_L^* = J_L/2\xi_k$ introduced into the program differs from that indicated by (1.55) by the factor $1/2\xi_k$.

to $\tilde{V}_k(0)$. On this basis we choose the initial point $\hat{z}$ of the integration of Eq. (2.3) for any fixed values of p and ξ.

We first seek a point $z = \bar{z}$ to the right of which the solution of (2.1) decreases monotonically in the mean square. It is known [11, 12] that this point is the largest root of the equation $p^2\rho(z) = \xi^2\mu(z)$. We take this root for the point $\bar{z}$.

We now seek a point $\bar{z}$ to the right of $\bar{z}$ such that the square-integrable solution $\tilde{V}(z)$ of Eq. (2.1) satisfies the condition

$$|\tilde{V}(\hat{z})| < \frac{|\tilde{V}(\bar{z})|}{N} . \tag{2.7}$$

where N depends on the required accuracy of the solution.

It can be proved that, if the function $b^2(z)$ is monotonic, then when the solution V(z) of the equation

$$V' = -\sqrt{\xi^2 - \frac{p^2}{b^2(z)}} \, V \tag{2.8}$$

satisfies the condition (2.7) the exponentially decreasing solution $\tilde{V}(z)$ of (2.1) satisfies the condition (2.7). Consequently the initial point $z = \hat{z}$ of the integration of Eq. (2.3) can be determined as follows: We integrate Eq. (2.8) from left to right from the point $\bar{z}$ with the initial condition $V(\bar{z}) = 1$ until we reach a point at which $V \leq 1/N$; we take this point for $\hat{z}$.

We integrate (2.3) by using the three-term Runge-Kutta formula with variable integration interval. For the integration from $z = \hat{z}$ to $z = \bar{z}$ we use the constant interval $h_1 = \hat{z}/N_1$, where the value of N_1 is chosen to yield the accuracy required of the solution. From $\bar{z}$ to 0 we use a variable interval h such that the increment $\Delta\theta$ of $\theta(z)$ when the variable changes from z to $z - h$ satisfies the condition $|\Delta\theta| \leq \pi/n$, where n depends on the accuracy required of the solution.

We denote the right-hand side of Eq. (2.3) by $f(z, \theta)$, i.e., we write (2.3)

$$\theta' = f(z, \theta). \tag{2.9}$$

The interval h at each point z is given by the formula

$$h = \min \left(\frac{\pi/n}{|f(z, \theta)|}, \frac{\hat{z}}{N_1} \right). \tag{2.10}$$

4. The Calculation of the Characteristic Values

The characteristic value ξ_k^2 is a root of the equation $\theta(\xi^2, 0) = 0$, where $\theta(\xi^2, \hat{z})$ is the solution of Eq. (2.3) with initial condition

$$\theta(\xi^2, \hat{z}) = -\arctan\left(\mu(\hat{z}) \sqrt{\xi^2 - \frac{p^2}{(b^2\hat{z})}} \right) - (k - 1)\pi.$$

To calculate this root we use a combination of Newton's method and the bisection method.

The initial approximation for $\xi_k^2(p)$ with a given p is obtained by parabolic extrapolation of the values of $\xi_k^2(p)$ for three previous values of p; the interval is constant relative to $T = 2\pi/p$ and not relative to p.

§3. The Basic Properties of Love Waves

In this section we present a qualitative description of Love waves; such a description can be useful for orientation or in the interpretation of observations. Fundamental information is given that is needed in the calculation of the dispersion and strength of Love waves by the program presented.

The properties of Love waves described in this section are direct consequences of characteristic-function theory [11, 23] or of Sections 1 and 2.

Some of the properties described are not rigorously demonstrated but are derived empirically—from large-scale calculations of the parameters of Love waves in various media; results of this type are specially mentioned in the text.

1. Definitions and Basic Formulas

At large horizontal distances from their source Love waves are determined by the following properties: They contain only horizontal transverse displacements (perpendicular to the direction of propagation or, what amounts to the same thing, to the azimuth at the source) and they are the main part of such displacements; the amplitude of a Love wave is proportional to the horizontal tangential component of the forces or stresses at the source.

Love waves can be decomposed into a series of harmonics (modes), the velocity of propagation and intensity of which depend on the vibration frequency.

The dependence of the phase and group velocities on the frequency or period is called the dispersion. The function giving this dependence for the k-th mode is called the k-th branch of the dispersion curve.

The amplitude spectrum of each separate mode has the form

$$U_\varphi^{(k)} = \frac{A \sin(\delta - \varphi)}{\sqrt{r}} \widetilde{F}(p, \, \xi_k) \overline{V}_k(p, \, z, \, h, \, \varphi). \tag{3.1}$$

Here and in the following k is the number of the mode ($k = 1,2,\ldots$); A is a constant; δ is the azimuth of the horizontal projection of the force; φ is the azimuth of the station; r is the distance from the epicenter; p is the circular frequency- $T = 2\pi/p$ is the group velocity; v_k is the phase velocity, $\xi_k = p/v_k$ is the wave number, c_k is the group velocity, $\widetilde{F}(p, \xi_k)$ is the space-time spectrum of the horizontal tangential component of the force at the source [formula (1.11)*]; $\overline{V}_k$ is the frequency characteristic of the "mean-source."

The function $\overline{V}_k$ is defined for an arbitrary source [formulas (1.53), (1.54)] satisfying the conditions given on page 2. For forces concentrated in the plane z = h and for displacements on the surface we have

$$\overline{V}_k(p, \, z, \, h) = \sqrt{\xi_k} \, \widetilde{V}_k(p, \, 0) \, \widetilde{V}_k(p, \, h) \tag{3.2}$$

[formulas (3.1), (1.57)].

In the case of a dipole with a moment we have

$$\overline{V}_{dk} = \sqrt{\xi_k} \, | \, [iE\widetilde{V}_k(p, \, 0) \widetilde{V}_k(p, \, h)\xi_k + G\widetilde{V}_k(p, \, 0)\widetilde{V}_k'(p, \, h)] \, | \tag{3.3}$$

[formulas (3.1), (1.58)]. Here $\widetilde{V}_k(p, h)$ and $\widetilde{V}_k'(p, h)$ are functions of p and h determining the intensity of the Love waves (Sec. 2); $E = \sin\gamma \cos(\alpha - \varphi)$; $G = \cos\gamma$; γ is the dip of the plane of discontinuity; α is the azimuth of the dip of the same plane; φ is the azimuth at the observation station.

---

*This and subsequent references to formulas of Sections 1 and 2 are given only for completeness. The reader who is not interested in the mathematics can ignore them.

Hence the relation between the mode characteristics for a concentrated force $\overline{V}_k$ and for a dipole $\overline{V}_{dk}$ is

$$\overline{V}_{dk} := \left| iE\, [\overline{V}_k]\, \xi_k + G\, \frac{d}{dz}\, [\overline{V}_k] \right|. \tag{3.4}$$

The dispersion curves $v_k(T)$ and $C_k(T)$, the frequency characteristics of the medium $\overline{V}_k$, and the theoretical seismograms of the various modes of Love waves can be calculated by using programs whose basic descriptions are given in Paragraph 9 of this section.

2. The Dispersion

The Phase Velocity v_k. Dispersion curves are of the form shown schematically in Fig. 1. For all modes the branches of dispersion curves have the initial value $v_k = b(Z + 0)$ — the maximum velocity of transverse waves in the medium (under the condition that this velocity corresponds to the maximum depth below which the medium is homogeneous).

The modes are numbered in order of increasing boundary frequency $p = p_{gk}$, i.e., the maximum possible frequency of vibrations in the mode under consideration. For the first mode $p_{g_1} = 0$, i.e., this mode exists for the whole range of frequencies.

Hence to any frequency p there corresponds one (v_1) or several (v_1, v_2,...) phase velocities corresponding to various modes. For sufficiently large p the number of simultaneously possible modes is arbitrarily large; the number of modes present (the amplitude of these modes having a maximum for a given p) can be very large. For a given v_k, the corresponding frequencies p_k increase with the number k.

For any given p we have $v_{k+1} > v_k$ or, in other words, the phase velocity increases with increasing k and the branches of the dispersion curves corresponding to different values of k never intersect (this refers only to phase velocities). When p increases v_k decreases monotonically, while for all k the asymptote v_k is the absolute minimum of b(z), independently of whether b(z) has local extrema or discontinuities.

The Group Velocity C_k. The group velocity $C_k = dp/d\xi_k$ is related to v_k by the equation

$$C_k = \frac{v_k}{1 - \dfrac{p}{v_k}\dfrac{dv_k}{dp}}. \tag{3.5}$$

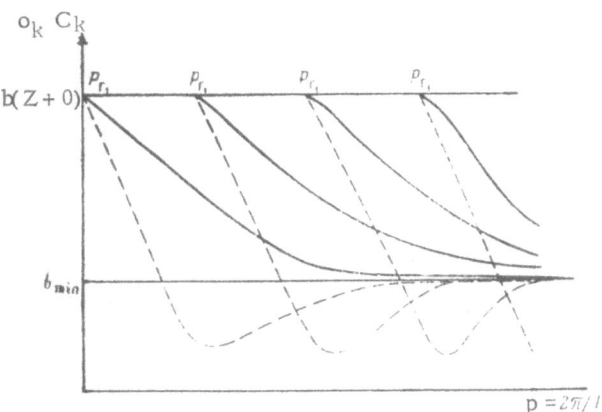

Fig. 1. Dispersion curves of the phase velocity (continuous curves) and of the group velocity (dotted curves).

The curves $C_k(p)$ exist in the same intervals of p for which $v_k(p)$ exists. It is clear from (3.5) and Fig. 1 that $C_k \le v_k$ and that the difference between C_k and v_k becomes greater when the curve for $v_k(p)$ becomes steeper. Just as for v_k we have $C_k \to b(Z + 0)$ when $p \to p_{g_k}$ and $C_k \to \min b(z)$ when $p \to \infty$. In contrast to v_k, however, the decrease of C_k with increasing p need not be monotonic, i.e., there are local extrema for which $d^2\xi/dp^2 = 0$ [In some intervals C_k can be smaller than min b(z).] To each extremum there correspond specific vibrations—the Airy phases decribed on pages 18 and 19.

The number of extrema of $C_k(p)$ can be different for different k, but this number cannot decrease with increasing k; it usually increases. With increasing k the fall in velocity between adjacent extrema increases. There can be various relations between C_{k+1} and C_k for different values of p; the curves $C_K(p)$ for different k can intersect.

3. Theoretical Seismograms

A method of constructing approximate theoretical seismograms (for large r) [described in formulas (1.63)-(1.65)] follows: The group velocity curve is decomposed into a series of monotonic sections between adjacent extrema. To each section there corresponds a quasisinusoidal vibration of variable frequency and amplitude. At time t the frequency p is determined by the condition

$$C_k(p) = r/t, \tag{3.6}$$

and the amplitude is given by the formula

$$\frac{A\overline{V}_k(p)\,\widehat{F}(p)}{r\sqrt{\left|\dfrac{d^2(p/v_k)}{dp^2}\right|}} \tag{3.7}$$

[formula (1.63)]. We can introduce a factor of the form $\exp[-\alpha_K(p)r]$ in the amplitude to give an approximate estimate of the effect of attenuation.

Hence for those sections for which the group velocity decreases with increasing frequency, the frequency on the record increases with time and conversely.

To obtain the theoretical seismogram for a given mode, we must add the vibrations corresponding to the different sections of the group-velocity curve. Here vibrations with different frequencies but with the same group velocity will interfere with one another, so that the seismogram can be rather complex. To the vicinity of extrema $C_K(p)$ there will correspond beats — the superposition of vibrations with frequencies that differ only slightly.

Each extremum of $C_K(p)$ yields a contribution to the theoretical seismogram; this is the so-called Airy phase — a vibration of constant frequency $\overline{p}$ corresponding to a given extremum of $C_K(p)$ that is damped with time [formula (1.66)].

The amplitude in the Airy phase at time t is obtained from the formula

$$\frac{A\overline{v}_k(\overline{p})\,\widehat{F}(\overline{p})v(\tau)}{\left(\dfrac{r}{2}\right)^{5/6}\sqrt[3]{\left|\dfrac{d^3(p/v_k)}{dp^3}\right|_{p=\overline{p}}}}, \tag{3.8}$$

where $v(\tau)$ is the Airy function with the argument

$$\tau = \frac{t - \dfrac{r}{C_k(\overline{p})}}{\sqrt[3]{\dfrac{r}{2}\left|\dfrac{d^3(p/v_k)}{dp^3}\right|_{p=\overline{p}}}} \tag{3.9}$$

given in Table 1. For positive τ, the function $v(\tau)$ decreases monotonically with increasing t. There is a more complete table of $v(\tau)$ in [24]. Examples of the calculation of Airy phases are given below (pp. 42-43 and 61).

Table 1

τ	v	τ	v	τ	v	τ	v	τ	v
−2.00	0.4031	0.00	0.6293	2.00	$0.6190\cdot10^{-1}$	4.00	$0.16866\cdot10^{-2}$	6.00	$0.17632\cdot10^{-4}$
−1.80	0.6040	0.20	0.5383	2.20	$0.4539\cdot10^{-1}$	4.20	$0.11122\cdot10^{-2}$	6.20	$0.10675\cdot10^{-4}$
−1.60	0.7619	0.40	0.4515	2.40	$0.3289\cdot10^{-1}$	4.40	$0.7267\cdot10^{-3}$	6.40	$0.6412\cdot10^{-5}$
−1.40	0.8715	0.60	0.3719	2.60	$0.2355\cdot10^{-1}$	4.60	$0.4705\cdot10^{-3}$	6.60	$0.3822\cdot10^{-5}$
−1.20	0.9327	0.80	0.3010	2.80	$0.16680\cdot10^{-1}$	4.80	$0.3019\cdot10^{-3}$	6.80	$0.2261\cdot10^{-5}$
−1.00	0.9493	1.00	0.2398	3.00	$0.11683\cdot10^{-1}$	5.00	$0.19204\cdot10^{-3}$	7.00	$0.13279\cdot10^{-5}$
−0.80	0.9280	1.20	0.18810	3.20	$0.8096\cdot10^{-2}$	5.20	$0.12111\cdot10^{-3}$	7.20	$0.7741\cdot10^{-6}$
−0.60	0.8771	1.40	0.14541	3.40	$0.5552\cdot10^{-2}$	5.40	$0.7574\cdot10^{-4}$	7.40	$0.4479\cdot10^{-6}$
−0.40	0.8645	1.60	0.11084	3.60	$0.3769\cdot10^{-2}$	5.60	$0.4697\cdot10^{-4}$	7.60	$0.2574\cdot10^{-6}$
−0.20	0.7201	1.80	0.08337	3.80	$0.2534\cdot10^{-2}$	5.80	$0.2889\cdot10^{-4}$	7.80	$0.14681\cdot10^{-6}$

The complete theoretical seismograph is, of course, the sum of vibrations in all modes.

In conclusion we note that, even when the frequency characteristics $\overline{V}_k$ of the medium are unknown, we can obtain a rather complete representation of the theoretical seismogram from a single group-velocity curve. For example the record begins at time $t_n = [r/\max C_K(p)]^{-t'}$ and ends at time $t_k = [r/\min C_K(p)]^{+t''}$. The lead t' is related to the maximum and t" to the minimum of the velocity. The duration of the Airy phase increases with increasing distance r and with decreasing radius of curvature of the curve $C_K(p)$ at the extremum. If we wish to take account of the distortion due to the apparatus, we must consider the maximum and minimum of $C_K(p)$ within the limits of the recorded part of the spectrum. Because of known effects related to the truncation of the spectrum, the recording interval will be somewhat broadened.

It is also clear from (3.6) that when the group-velocity curve is steeper the duration of the record of a single period will be longer, etc.

Theoretical seismograms are needed, in particular, for the clarification of conditions for recording, and as criteria for the detection of separate modes.

4. Scaling Laws

The following scaling laws hold for v_k and C_k:

a) When the transverse-velocity and depth units are multiplied by the same factor L with p fixed, the velocities v_k and C_k are multiplied by the same factor:

$$v_k(p)_{[b(z),\rho(z)]} = Lv_k(p)_{[b/L(Lz),\rho(Lz)]} \tag{3.10}$$

with a similar relation for C_k; here and in the following the corresponding model of the medium is given by the expression in square brackets;

b) When there is a change of scale in the depth and a proportional change of scale in the time (corre - sponding to an inversely proportional change in the frequency), the velocities v_k and C_k are unchanged:

$$v_k(p)_{[b(z),\,\rho(z)]} = v_k(p/L)_{[b(Lz),\,\rho(Lz)]} \tag{3.11}$$

with a similar relation for C_k.

Scaling laws analogous to (3.10) and (3.11) also hold for the frequency characteristics $\overline{V}_k$:

$$\overline{V}_k(p,\ \xi_k,\ z,\ h)_{[b(z),\,\rho(z)]} = L^q\overline{V}_k(p,\ \xi_k/L,\ Lz,\ Lh)_{[b/L(Lz),\rho(Lz)]}, \tag{3.12}$$

$$\overline{V}_k(p, \xi_k, z, h)_{[b(z), \rho(z)]} = L^s \overline{V}_k(p/L, \xi_k/L, Lz, Lh)_{[b(Lz), \rho(Lz)]},$$ (3.13)

where q and s are constants.

In the same way, a scaling law for the amplitude μ_φ^k will hold if the origins are the same; i.e., if

$$F(x, y, z, t) = L^j F(Lx, Ly, Lz, Lt),$$ (3.14)

where j is a derived constant.

5. The Principle of Reciprocity

It follows from (3.1) that the intensity U_φ^k is proportional to some function $\overline{V}(p, z, h)$. For a force concentrated at a point we have

$$\overline{V}_k(p, z, h) = \overline{V}_k(p, h, z)$$ (3.15)

[formula (1.57)].

For the following it is sufficient to consider the case in which the point of observation is on the surface of the medium (z = 0) and the depth of the source is altered by a constant factor h. In other words we consider $\overline{V}_k(p, h, 0)$; from it we can obtain the value of $\overline{V}_k$ when $z \neq 0$ from the following formula which is a consequence of (3.2) and (3.15):

$$\overline{V}_k(p, h, z) = \overline{V}_k(p, h, 0) \frac{\overline{V}_k(p, z, 0)}{\overline{V}_k(p, 0, 0)}.$$ (3.16)

6. The Variation of Intensity with Depth

The dependence of $\overline{V}_k$ on the depth z is different in the intervals $0 < z < \overline{z}_k$ and $z \geq \overline{z}_k$. Here $\overline{z}_k$ is the maximum of those depths z for which (for fixed p) the inequality $b(z) \geq v_k(p)$ holds. The quantity $\overline{z}_k$ is the depth of the point of inflection of the function $\overline{V}_k(z)$ for given p, and of course for fixed $v_k(p)$ [formula (2.1)]. The definition of $\overline{z}_k$ is illustrated in Fig. 2a. Above this point $\overline{V}_k(z)$ is an oscillating function of z and has k − 1 zeros (nodes). The behavior of $\overline{V}_k(z)$ in the interval $0 \leq z \leq \overline{z}_k$ is considered for concrete examples in Sections 4, 7 and 10. Below the points $z = \overline{z}_k$ the function $\overline{V}_k(z)$ decreases monotonically as z increases with an approximately exponential rate of decrease:

$$\overline{V}_k(z) \sim \exp\left\{-\frac{p}{v_k} \int_{\overline{z}_k}^{z} \sqrt{1 - \left[\frac{v_k}{b(z)}\right]^2}\, dz\right\}.$$ (3.17)

Hence $z = \overline{z}_k$ is approximately the depth of penetration of the k-th mode of a Love wave. Examples of graphs of $v_k(z)$ (k = 2) are shown for various values of p in Fig. 2b. When z reaches the value Z (the depth at which b and ρ become constant), the rate of damping of $\overline{V}_k(z)$ with increasing z becomes strictly exponential.

In the profiles encountered in practice b(z) increases with depth z over most of the interval $0 < z < Z$. On the other hand $v_k(p)$ decreases with increasing p. It thus follows that the depth of penetration $\overline{z}_k$ decreases when the frequency p increases, and the drop of $|\tilde{V}_k(z)|$ below this point becomes less pronounced.

If there is a boundary in the medium at depth H at which b(z) increases discontinuously, the spectrum of a Love wave decomposes into two parts. The higher-frequency part is defined by the condition $v_k(p) < b(H+0)$, so that $z_k \leq H$ in this part. Vibrations with these frequencies are mainly propagated above the boundary and depend weakly on the properties of the medium below the boundary. For the lower-frequency part $v_k(p) > b(h+0)$ which means that $\overline{z}_k > H$; Love waves with these frequencies penetrate below the boundary and their properties are mainly determined by the structure of the lower medium.

For a fixed frequency, an increase in the number of the mode leads to an increase in the depth $\overline{z}_k$ [since $v_k(p)$ increases]; in other words the penetration is greater when the number of the mode is greater. This effect is illustrated in Fig. 2c.

For fixed velocity v_k, the depth $\overline{z}_k$ is by definition independent of k. On the other hand when k is larger, then the value of p_k corresponding to the given v_k is larger. Hence [formula (3.17)] if $b(z)$ increases with z the damping of $\overline{V}_k(z)$ with depth for $z > \overline{z}_k$ will be more rapid when k is larger.

Dipoles. For dipoles with a moment [formulas (1.58), (3.3), (3.4)], and other dipoles, the variation of the intensity with depth depends not only on $\overline{V}_k(z)$, but also on $d\overline{V}_k(z)/dz$. * We note that for $z = 0, d\overline{V}_k(z)/dz = 0$. Since the zeros of $d\overline{V}_k(z)/dz$ coincide with the extrema of $\overline{V}_k(z)$, the variation with depth of the intensity of Love waves above the point of inflection ($z < \overline{z}_k$), and thus the frequency characteristic of the medium in this region, can basically depend on the parameters of the dipole. For $z > \overline{z}_k$ the intensity monotonically decreases with z approximately exponentially as before. This problem is considered in more detail in Section 10.

7. The Dependence of the Intensity on the Frequency and the Mode Number

We do not give any theoretical estimate of the dependence of $\overline{V}_k$ on p and k. However analysis of a set of calculations for various models of the earth's crust and the upper mantle yield the following rules.

a) When $p \to \infty (T \to 0)$, $\overline{V}_k$ tends asymptotically to zero. At the beginning of a branch ($p = p_{g_k}$), $\overline{V}_k(z) = 0$.

The variation of $\overline{V}_k$ with p depends essentially on the depth h of the source. For $h > Z$, the graph of $\overline{V}_k(p)$ is a smooth curve without zeros. For $0 < h < Z$ the frequency characteristic for the k-th mode can have zeros, the number of which decreases with h. For $h = 0$, $\overline{V}_k(p)$ has no zeros, but it has one or several extrema for approximately the same frequencies for which the group velocity C_k has minima. The maxima of $\overline{V}_k(p)$ correspond to the minima of C_k and conversely.

b) The damping of $\overline{V}_k$ with increasing p is slowest close to minima of the velocity $b(z)$. With increasing p, Love waves become increasingly concentrated in a narrow neighborhood of these minima. This concentration proceeds differently for absolute and local minima.† All the modes become concentrated at an absolute minimum; when the mode number decreases the concentration begins with smaller p. At local minima the concentration proceeds by the successive replacement of one mode by another, starting with some k not necessarily equal to one (see below, paragraph 8).

c) The dependence of $\overline{V}_k$ on k for fixed p differs essentially for different z, because of the vibrational character of $\overline{V}_k(z)$ for $z < \overline{z}_k$ (different for different values of k). It is thus useful to compare the minimum or mean values of $|\overline{V}_k(z)|$ for the whole range of z. The relation $|\overline{V}_k(z)| \leq |\overline{V}_k(0)|$ always holds in a medium with monotonically increasing $b(z)$. If on the other hand there is an internal minimum of $b(z)$ in the medium, this relation can be violated in certain intervals of p because of the concentration of vibrations in a waveguide.

---

*In general, the displacement in Love waves due to multipoles of order n is a linear combinations of expressions of the type

$$(p/v_k)^{(n-j-1)} d^j \overline{V}_k(z)/dz^j \quad (j = 1, 2 \ldots n-1).$$

†An absolute minimum occurs most often at the surface of a layer; the zones of minima of $b(z)$ at a certain depth are usually called waveguides.

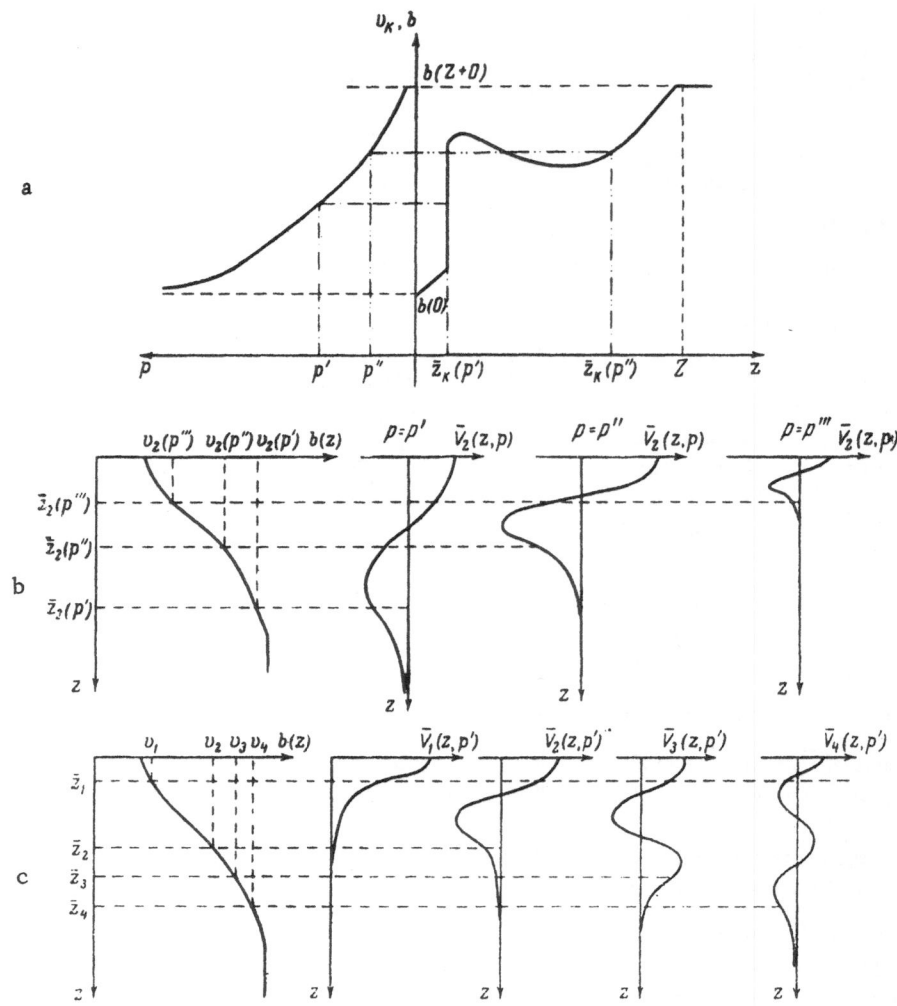

Fig. 2. The variation of the intensity $\overline{V}_k$ with the depth z.   a) The determination of the depth of penetration of the k-th mode: to the right, the velocity profile b(z); to the left, the dispersion curve $v_k(p)$; b) examples of graphs of $\overline{V}_2(z)$
for three different frequencies ($p' < p'' < p'''$), to the left, the velocity profile
b(z); c) examples of graphs of $\overline{V}_k(z)$ for fixed frequency $p = p'$ (k = 1, 2, 3, 4),
to the left, the velocity profile b(z).

Here we consider only the first case. As Fig. 3 shows schematically, $\overline{V}_k(0)$ decreases with increasing k.
In particular $|\overline{V}_1(0)|$ is larger than all other $|\overline{V}_k(0)|$ for any p. However, the decrease of max $|\overline{V}_k(z)|$ with k is
not monotonic and is very slow. Thus for some models, such as thick layers with a weak positive velocity
gradient, max $|\overline{V}_3(z)|$ and max $|\overline{V}_{20}(z)|$ differ by less than one order in a definite interval of p.

Clearer rules are obtained if we compare max $|\overline{V}_k(p, z, 0)|$; i.e., if we take the maximum frequency
characteristics relative to p for various values of z and choose the maximum of these maxima. These maxima
generally decrease with k, but they can occur for different frequencies increasing with k. Hence in a given
frequency interval, if we exclude the mode with k = 1, certain modes, but not necessarily the lowest, will have
the greatest intensity; when this frequency interval is varied in the direction of higher frequencies the numbers
of these harmonics will increase.

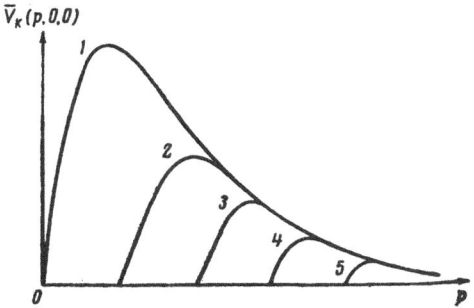

$\overline{V}_K(p, 0, 0)$

Fig. 3. Frequency characteristics $\overline{V}_K(p, 0, 0)$ for a surface source and a surface receiver (schematic): The numbers on the curves are the mode numbers.

For sufficiently high frequencies it thus follows that it is impossible to represent the over-all field of SH surface waves by the superposition of a few low modes. This refers in particular to the case of a very deep focus, since higher modes penetrate much more deeply into the medium than lower modes.

The existence of some tens of modes, all of which are important because of their intensity, contradicts the traditional representation. (It is possible that most of them combine into short vibrations in the group of "body" waves, as occurs for the spherical model [61, 62].) A quantitative solution of this interesting problem has, however, not yet been obtained.

8. Love Waves in Waveguides

The propagation of Love waves in media in which $b(z)$ has several minima has some interesting features. For sufficiently large p, SH waves are concentrated in a narrow neighborhood of that minimum at which a source is located. For intermediate values of p, simultaneous generation and interference of waves can occur in the vicinity of some minima of $b(z)$. For small values of p, the presence of local minima is not noticeable. These general rules are different for different modes.

Phase Velocities. In Fig. 4 we show schematic dispersion curves for the case when, in addition to an absolute minimum $b(0)$ at the surface, there is an internal minimum $b(\tilde{z})$ at the point $z = \tilde{z}$ (a waveguide). We see that the curves for $v_K(p)$, starting at a certain value of k, become similar to step-functions — there are alternating intervals of slow and rapid decreases of v_k; the number of "steps" in a branch increases with increasing k. The successive curves for $v_K(p)$ fit into each other more and more closely at the ends of the "steps" when k increases. The flat parts of the branches appear to be continuations of one another, and together they form a system of discontinuous curves $v_N^*(p)$ (N = 1, 2, . . .), the combination of which tends to $b(\tilde{z})$ when p increases. For small p, these curves are bounded by certain values p_N which increase with N (if N increases upwards) and which depend on $b(z)$ in the vicinity of a local minimum of the velocity. In the interval $p_N < p < \infty$, any curve $v_N^*(p)$ practically coincides with the curve $v_N(p)$ for the N-th mode in models differing from those under consideration in the absence of a minimum of $b(z)$ at the surface, so that the single minimum of $b(z)$ is at the point $\tilde{z}$ (see Fig. 4a).

The value $k = k^*$ starting with which the branches of $v_K(p)$ have "steps" depends mainly on the structure of the medium in the zones of local and fundamental minima of $b(z)$; in fact k^* will be smaller when the local minimum is sharper, when $\tilde{z}$ is smaller, and finally when the region of fundamental minimum is more weakly expressed. In the limit $k^* = 1$, in contrast to the other curves, the curve for $v_1(p)$ can have a step that begins with a smooth transition, i.e., it has no clear-cut low-frequency boundary.

Group Velocities. Figure 4c shows that, to the sloping sections of the curves for $v_K(p)$, there correspond sloping sections of the curves for $C_K(p)$ that pass only a little below the ordinate $C = b(\tilde{z})$ [this follows from (3.5), since on the sloping sections of $v_K(p)$ the derivative dv_k/dp is small]. In the region where $v_K(p)$ decreases sharply ($|dv_k/dp|$ is large) there are minima of $C_K(p)$. The main minimum of $C_K(p)$ corresponds to the highest-frequency sharp drop of $v_K(p)$ before its passage towards the asymptote $b(0)$. It also exists in models without a local minimum of $b(z)$ (see Fig. 1).

The position and configuration of these minima depend on the number of the mode k. As k increases the number of local minima of $C_K(p)$ increases; the frequencies corresponding to the minima increase; the widths of the minima decrease and their depths increase. For fixed k a similar effect results from a change in the properties of the model, i.e., from a strengthening of the local minimum or a weakening of the fundamental minimum of b(z). The existence of sharp local minima of $C_k(p)$ separated by sections where $C_k(p) \approx b(\tilde{z})$, is a distinguishing criterion of models with waveguides.

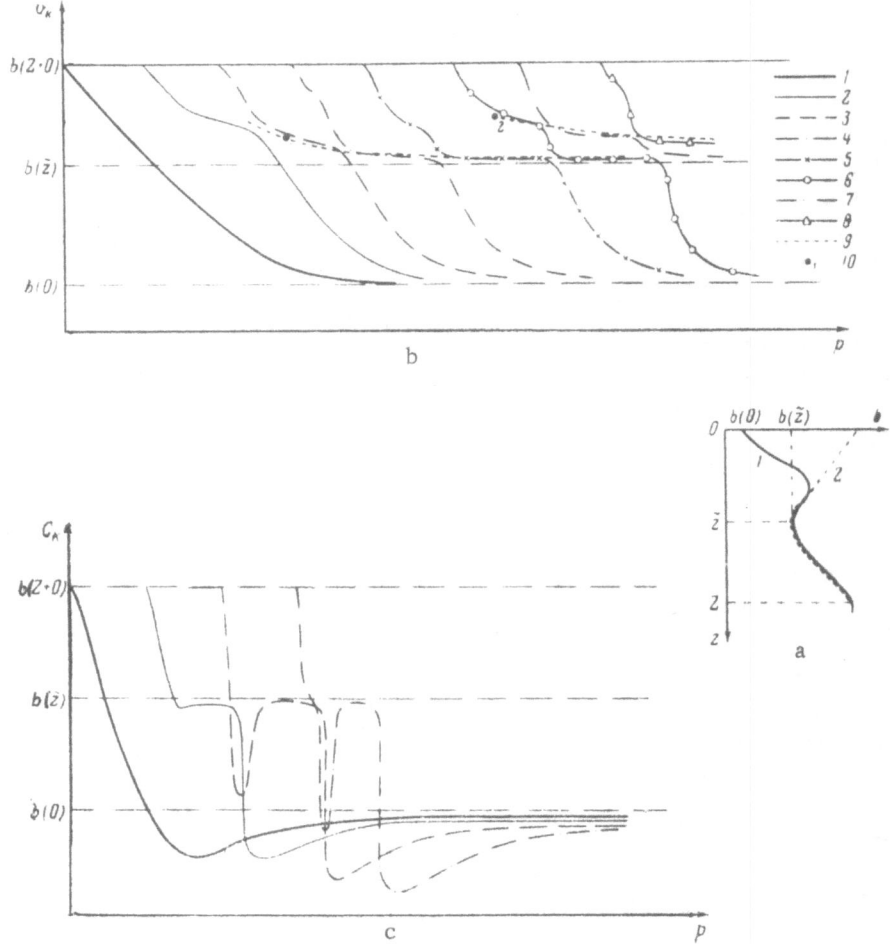

Fig. 4. Schematic graphs showing the dispersion of Love waves in a medium with a waveguide. a) Models of media with waveguides: continuous curve, the absolute minimum of $b(z)$ at the surface (model 1); dots, the absolute minimum of $b(z)$ in the waveguide (model 2); $\tilde{z}$, depth of minimum point of $b(z)$ in waveguide; b) phase-velocity dispersion; c) group-velocity dispersion; in b and c, 1-8 are the first eight modes in model 1; 9 is the first mode and second node in model 2; 10 gives the boundary frequencies $\dot{p}_N$ of Love waves in model 2 (channel waves).

It follows from Paragraph 3 that specific waves with quasistationary periods must correspond to narrow minima of $C_k(p)$. These waves arise only in media containing waveguides. The approach of the branch $v_k(p)$ to the branches $v_{k \pm 1}(p)$ at the ends of "steps" leads to the intersection of the corresponding curves for $C_k(p)$ with those for $C_{k \pm 1}(p)$; at the same frequencies for which the curve $C_k(p)$ has a local minimum, the curve $C_{k-1}(p)$ falls sharply towards its fundamental or local minimum. It thus follows that the separation of specific waves related to local minima of $C_k(p)$ can be hindered by waves belonging to the preceding modes and not necessarily related to the local minimum of the velocity $b(z)$.

Specific examples of this phenomenon will be considered in Sections 8 and 9.

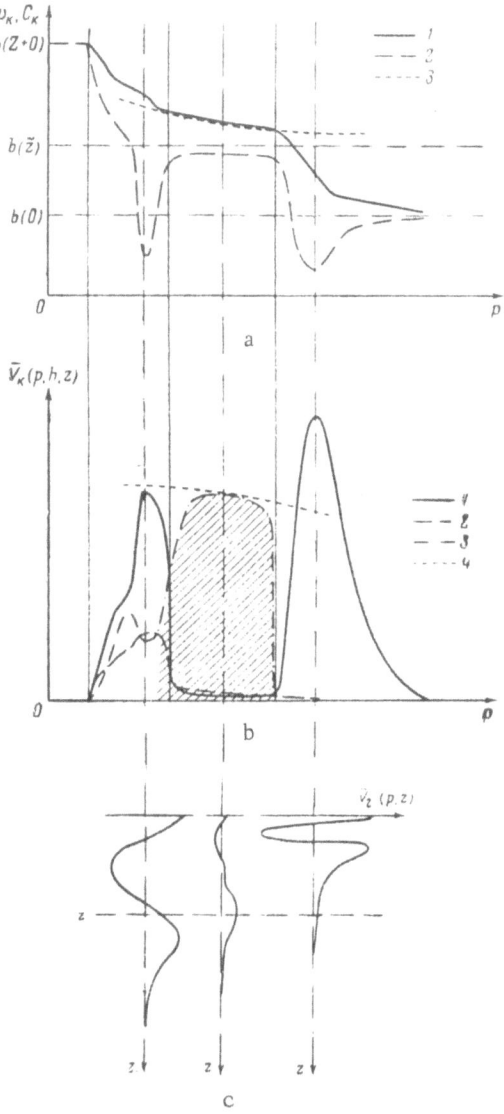

Fig. 5. Schematic graphs of the disturbances due to Love waves
in a medium containing a waveguide (model 1 of Fig. 4a).
a) Dispersion of the k-th mode: 1) phase velocity; 2) group
velocity; 3) phase velocity $v_1^*(p)$ in model 2 of Fig. 4a; b) frequency characteristics of the medium: 1-3) k-th mode in model
1 of Fig. 4a; 1) source and receiver on the surface; 2) source
in the waveguide $(h = \tilde{z})$, receiver on the surface; 3) source and
receiver in the waveguide '$(h = z = \hat{z})$ (the dotted part of the spectrum corresponding to a channel wave); 4) first mode of model 2
in Fig. 4a, source and receiver in the waveguide $(h = z = \tilde{z})$;
c) variation of the intensity $\overline{V}_k$ with the depth z for various sections of the spectrum of the k-th mode. For definiteness we take
k = 2; the horizontal scale of the mean curve is greatly ex-
aggerated.

Intensity. Figure 5 shows schematically the disturbance of isolated modes in a medium containing a waveguide. Starting with some $k = k*$, each mode is concentrated in the vicinity of a local minimum of $b(z)$, but only at the ends (one or several) of frequency intervals corresponding to sloping sections of the dispersion curve $v_K(p)$. To each section corresponds one or several local extrema of $\overline{V}_K(z)$ with respect to z inside the waveguide. Love waves concentrated in the vicinity of local minima of $b(z)$ are usually called SH channel waves. Each discontinuous curve $v_N^*(p)$ corresponds to a channel wave of the N-th order (i.e., there are $N-1$ displacement nodes in the waveguide). Its spectrum is a combination of those parts of the spectra of the different modes which correspond to the discontinuous curve for v_N^* in Fig. 4b. It is obvious that this spectrum is bounded below by p_N: For $p < p_N$ concentration of vibrations in the waveguide practically ceases (Fig. 5b).

We see that the conditions under which channel waves can be observed are very restrictive: for small p such waves do not occur; for large p they are rapidly damped out with increasing distance from the waveguide and they can be observed only if the depth of the waveguide is sufficiently small.

It will be shown below (p.61) that channel waves in the waveguides that are sometimes said to exist in the mantle cannot be observed in practice at the surface of the continental crust.

Another interesting feature of the behavior of $\overline{V}_K(p)$ in media containing waveguides is the existence of sharp local maxima of $\overline{V}_K(p)$ at frequencies corresponding to local minima of the curves for $C_K(p)$. These maxima are observed for a wide range of depths (from the surface to the waveguide inclusive). As has already been noted, they correspond to specific waves; these waves will be investigated in more detail in Section 9.

9. A Program for the Calculation of Dispersion and Intensity

Parameters of the Medium. The velocity $b(z)$ of transverse waves and the density $\rho(z)$ are piecewise-broken functions. They can be defined by the sequence of constants

$$b(z_0), \ b(z_1 - 0), \ b(z_1 + 0), \ b(z_2 - 0), \dots, b(z_m - 0), \ b(z_m + 0),$$
$$\rho(z_0), \ \rho(z_1 - 0), \ \rho(z_1 + 0), \ \rho(z_2 - 0), \dots, \rho(z_m - 0), \ \rho(z_m + 0),$$
$$z_0, \qquad z_1, \qquad z_1, \qquad z_2, \dots, \qquad z_m, \qquad z_m = Z.$$

If $b(z_j - 0) = b(z_j + 0)$ and $\rho(z_j - 0) = \rho(z_j + 0)$, i.e., the velocity and density are continuous at $z = z_j$, then for this z_j only a single value of $b(z_j)$, $\rho(z_j)$ and z_j need be given. The total number of points (counting a point of discontinuity as two points) must not exceed 128.

Computation Parameters. Here we have the upper values (k_u, T_u, v_u) and lower values (k_l, T_l, v_l) of the relevant mode number, the number of periods, and the phase velocity. The calculations are performed for $k_l \leq k \leq k_u$, $T_l \leq T \leq T_u$, $v_l \leq v \leq v_u$. From the discussion on p.17, we have $v_u \leq \max b(z)$, $v_l \geq \min b(z)$. In addition to these quantities, the constant interval ΔT between adjacent points at which calculations are performed, determining the detail that is desired, must be specified. From Paragraph 2, more detail is usually needed for higher modes than for the fundamental mode.

Extra parameters are the constants determining the position or density of points on the z axis for which the values of $\widetilde{V}_K(0)\widetilde{V}_K(z)$ are printed.

Output Data (Printed). The following information is printed for each mode in the interval $k_l \le \overline{k \le k_u}$:

a) the boundary values T^* and p^* corresponding to $v_k = v_u$ (if $T^* < T_u$);

b) for each T_i in the range $T_l \le T \le T^*$ (with given interval ΔT): T_i, p_i, $v_k \dagger$ C' and the values of $\tilde{V}_k(0)$ $\tilde{V}_k(z_s)$, $\tilde{V}_k(0)$ $\tilde{V}_k(z_s)$ at given points z_s or with the given z interval ‡

c) a sequence T_i, p_i, v_k, C_k, and $(\,|d^2 \xi_k/dp^2\,|_{p \,=\, p_i})$ in the required T range.

---

†The values of v_k are printed twice and correspond to different methods of calculation. Their agreement serves as a rather reliable check of machine operation, as each value of v_k is calculated independently of the previous calculations. The only other check is a periodic verification of the constancy of the cyclic sum in the program and the original data.

‡The values of V_k and V_k' are normed by using the factor $(J_L^*)^{1/2}$.

CHAPTER II

LOVE WAVES IN THE EARTH'S CRUST

In this chapter we consider calculated intensities and dispersions of several lower modes of Love waves in the period range in which these waves are propagated mainly in the earth's crust and in which they practically never penetrate into the mantle (i.e., their phase velocities are lower than the velocities of transverse waves in the mantle — see pp. 20-21).

In Section 4 we investigate frequency characteristics for various models of the earth's crust. In Section 5 dispersion is studied for the same models and the possibility of determining the structure of the crust on the basis of Love waves is estimated. In Section 6 we investigate the conditions under which it is possible to record various modes, and we describe theoretical seismograms for some models of the medium.

§4. Frequency Characteristics of the Medium

In this section we investigate the frequency characteristics calculated by means of the formulas (3.2)-(3.3) of Section 3 for a concentrated force and for variously oriented dipoles. The models of the continental crust shown in Fig. 6 were considered; they include a sufficiently wide class of hypotheses. The first eight models (Fig. 6A) are taken from [22]; the thickness H of the crust is assumed to be 30 km and the velocity at the surface to be $b(0) = 3.6$ km/sec.* Other models are obtained from the preceding models by the addition of surface layers with decreasing velocities. We consider three models of surface layers (Fig. 3B): a homogeneous layer with velocity $b = 2.5$ km/sec (model a); a layer with velocity increasing linearly from 2 to 3 km/sec (model b); a layer with velocity increasing from 2.5 km/sec to the value at the crystalline-crust surface (model c). The first two models correspond to a sedimentary crust, the third to a transitional (ruptured) layer at the top of crystalline crust. Hence there is a velocity discontinuity at the contact with the crystalline crust only on the first two models. For various calculations the thickness of the layers varies from 1 to 4 km. The surface layers a, b, or c are added to the basic models I-VIII so that the total thickness of the crust remains constant. The complete model is denoted by the combination of the indices in Fig. 6A and 6B (for example VI a).

In all the calculations, the Jeffreys-Bullen results for the velocity of transverse waves in the mantle [37] were used, and the Gutenberg value was used for a check (Fig. 6C, profiles 1 and 2). The effect of possible variations in the velocity profile of the mantle are considered in more detail in the following chapter. The density in the mantle is taken from Bullen's model A [37] (Fig. 6C). Calculations are for a concentrated force unless otherwise stated.

1. The Fundamental Mode; A Crust without a Surface Layer

Figure 7 shows the frequency characteristic[†] of the medium for the fundamental (first) mode corresponding to models II, VI, VII, VIII for three focal depths (h = 0, 20, 40 km). We see that the

---

*The main results are generalized to other values of H and b_0 by using scaling methods (see pp. 19-20 and also pp. 32-33 below).

[†]Here in the sequel the frequency characteristics are given for two-dimensional models; these differ from the characteristics for three-dimensional models only in the absence of the factor $\sqrt{\xi_k} = \sqrt{2\pi/Tv_k}$ [see formula (3.3)].

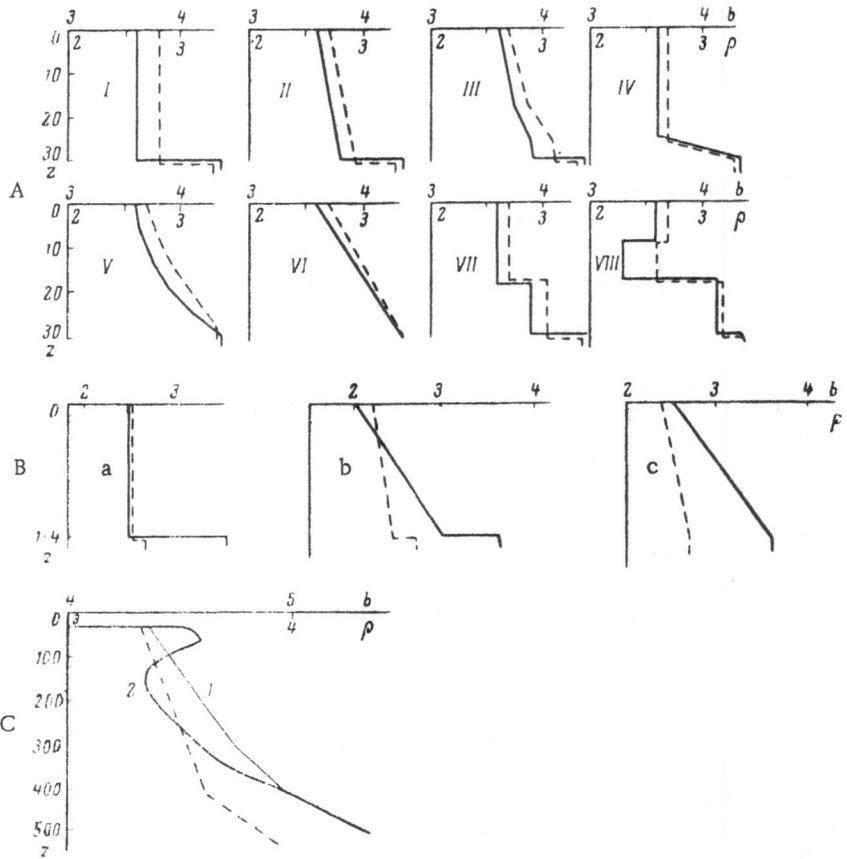

Fig. 6. Calculated models of the medium. Velocity b(z) — continuous curves (in km/sec); density $\rho(z)$ — dotted curves (in g/cm^3); A) crystalline earth's crust; Roman numerals — model index; B) surface layers, Latin letters — model index; C) mantle, Arabic numerals — model index; 1) according to Jeffreys-Bullen; 2) according to Gutenberg; density as in Bullen's model "A".

characteristics for the different models are in general similar. For all characteristics there is a smooth maximum in the 10-30 sec period range, which is displaced in the direction of larger T when h increases. For T > 50-60 sec, the characteristics for a given h for all models run together — the crustal structure does not influence the Love waves. For all models except those with a waveguide (VIII), $\overline{V}_1(h)$ decreases with increasing h for any values of T. The only exceptions are models with waveguides in the crust for some sufficiently small values of T (less than 5 sec) in those cases in which the source is above the axis of the waveguide (h < 15 km).

The main differences between frequency characteristics for different models are as follows.

a) When the transverse-wave velocity gradient increases, $\overline{V}_1(T, h)$ is damped more rapidly with decreasing period T (to the left of the maxima, of course) for h < H.

b) When the velocity gradient increases, $\overline{V}_1(T, h)$ is damped more rapidly with increasing depth h for T < 8 − 10 sec.

c) When the mean velocity in the crust increases, the maximum of $\overline{V}_1(T, h)$ is displaced in the direction of shorter periods.

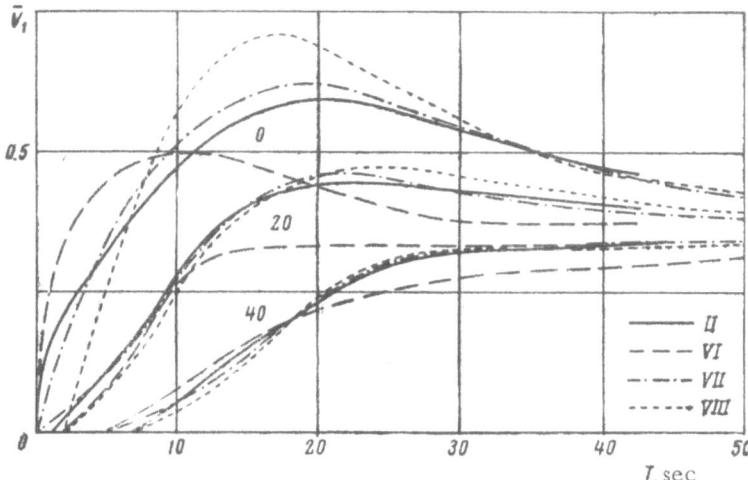

Fig. 7. Frequency characteristics of the medium $\overline{V}_1$, (T, h) for models
of the crust without a surface layer: II, VI, VII, VIII. Curve numbers
are focal depths (km).

d) A feature of one model with a wave-guide (VIII) is a very small value of $\overline{V}_1$(T, h) for T < 2 sec
even for h < H; in this period range Love waves are concentrated in the waveguide and never reach the surface
in practice.

A representation of frequency characteristics for earthquakes with intermediate-depth foci can be obtained from Fig. 20 [a continental crust with an inhomogeneous mantle containing a waveguide (2) or not
containing a waveguide (1)]. When the distance of the focus below the crust increases the frequency-charac-
teristic maximum is rapidly smoothed out, and conversely there is a steep drop of $\overline{V}_1$(T, h) to the left of the
period range 30-40 sec. This shows in particular that the sharp decrease in magnitude as determined from
surface waves and noted in [26] can be simply explained by a velocity discontinuity at the bottom of the crust,
and not by the presence of an asthenosphere, if the focus is at some depth below the crust.

2. The Fundamental Mode; a Crust with a Surface Layer

Concentrated Force. The effect of sedimentary or loose surface layers on the frequency characteristics of a medium corresponding to models IIb and IIc is shown in Fig. 8. It is plain from the graphs that a
surface layer with a low velocity leads to the appearance of a sharp resonance in the frequency characteristics
for small h at short periods (T ≤ 2 - 4 sec). When the focus is below the layer the resonance disappears and,
in contrast to models without surface layers, the frequency characteristic becomes fairly steep even for relatively small h (of the order of H/2 or 3H/2).

At boundaries of the type found at the bottom of sedimentary layers or at the bottom of the crust, the
velocity b(z) usually increases sharply. Calculations show that, for period ranges in which Love waves do not
penetrate greatly below such a boundary (see pp. 20-21), the medium below this boundary (core or mantle
respectively) can be replaced by a half-space with sufficient accuracy. Hence the scaling laws stated in
Paragraph 4, Section 3, Chapter I) are applicable in such a range. If these scaling laws are used, the frequency
characteristics given here can be easily applied in the case of models with other crust thicknesses (or sedi-
mentary-layer thicknesses) with the same ratios between the thicknesses of the different layers. To do this it is
sufficient to make an appropriate change of scale in T and to refer the calculated characteristics to a pro-
portionally changed focal depth. Thus for a surface focus (see Fig. 7), the doubling of the crust thickness in
model II (a crust of "rock systems", H ≈ 60 km) yields a wideband characteristic $\overline{V}_1$(T, 0) with maximum
amplitude for a period of about 40 sec instead of 20 sec. On the other hand if we decrease the thickness by a
factor of five (an oceanic crust) we obtain a very narrow-band characteristic with resonance for about 4 sec.

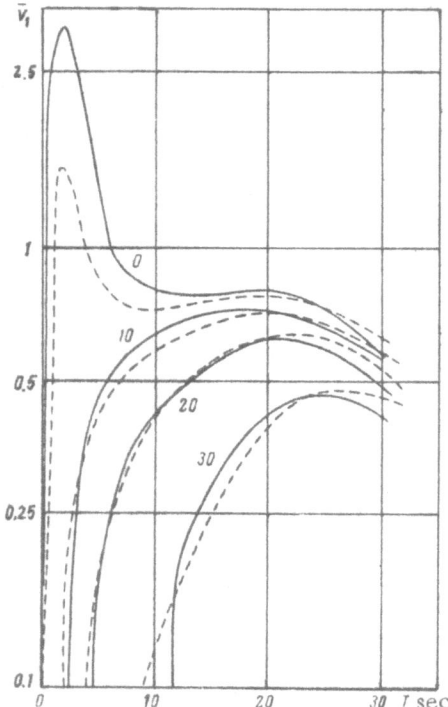

Fig. 8. Frequency characteristics of a medium $\bar{V}_1$ (T, h) for a model of the crust with a surface layer. Continuous curves) model IIb; dotted curves) model IIc; numbers on curves) focal depths (km).

By doubling the thickness of sedimentary rock in Model IIa we obtain a resonance of $\bar{V}_1(T, 0)$ that is less sharp at a period of 4 sec instead of 2 sec.

The addition above such models of a thin layer (h = 0.1-0.3 km) with a very low velocity (b = 0.3-0.5 km/sec, ρ = 1.8-2.0 g/cm³) corresponding to friable sediment leads (with a surface focus) to the appearance of an extra sharp peak in the frequency characteristic for very short periods T (less than 1 sec) (Fig. 9). This peak rapidly disappears as the depth of the source below the friable layer increases.

Dipoles. It follows from (3.4) that the frequency characteristics for a dipole and a concentrated force can differ considerably due to the appearance of the factor $\xi_k = p/v_k$ and an extra term proportional to (d/dh) ∇_k [T, h]. The role of this term in the resulting value of $\bar{V}_{dk}$ for a dipole depends essentially on the dipole parameters. Figure 10 shows graphs of V_{d_1} (T, h) for a dipole in model IIb; these graphs include a wide range of possible variation of the parameter E/G [depending, according to (3.3), on the angles γ and ($\alpha - \varphi$)]. A comparison of Fig. 8 with Fig. 10 shows that:

a) the transition from a simple force to a multipole leads to a decrease in the resonance periods;

b) the position of the maxima of $\bar{V}_{d_1}$ with respect to T for a dipole depends weakly on the orientation of the dipole;

c) the dependence of $\bar{V}_{d_1}$ on the source depth h for dipoles can differ considerably for different periods T and different values of E/G. At the limits of the crust there can be intervals of values where $V_{d_1}(T, h)$ increases with h. This problem is considered in more detail in Section 10, Chapter IV.

3. Higher Modes

Figure 11 shows examples of frequency characteristics for second and third modes in a homogeneous "continental" crust (model I) on a mantle of type 1 (see Fig. 6C, 1).

A comparison of Fig. 7 with Fig. 11 shows several differences between the higher modes and the fundamental.

1. For h < H the frequency characteristics have a clearly expressed main pass band — this band becomes narrower when the mode number increases. It corresponds to the propagation of vibrations in the limits of the crust; the penetration of waves into the mantle (see Section 7, Chapter III) corresponds to a very sharp drop of frequency characteristics at periods $T_2' = 9 - 11$ sec for the second mode and $T_2' = 5 - 6$ sec for the third mode.

2. As h increases, the steepness of the slope of the main pass band increases in the crust. The form of the curve for $\bar{V}_k(T)$ simultaneously becomes more complex; minima of $\bar{V}_k(T)$ appear, the positions of which vary with h.

3. When h increases further with h > H, the main pass band becomes narrower because of the weakening of the short periods for a fixed position of the right slope of $\bar{V}_k(T)$. The maximum value of $\bar{V}_k(T)$ for h > 200 km falls sharply; there is practically no penetration of the medium by vibrations in this period range. Hence for

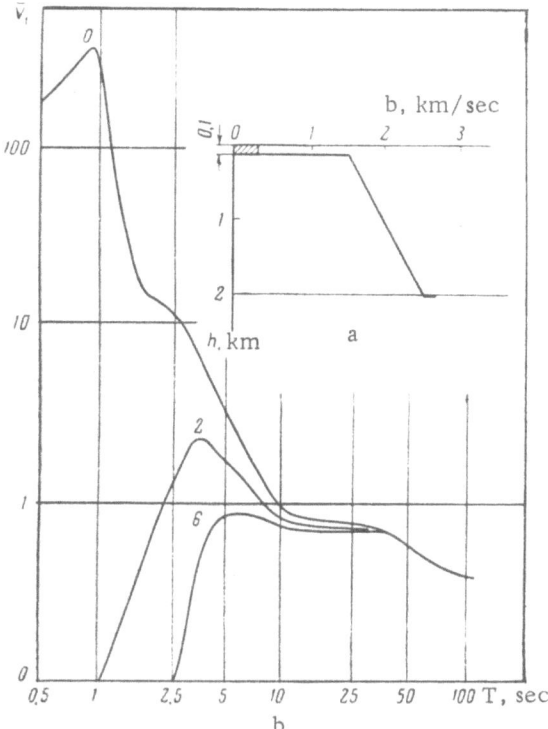

Fig. 9. The influence of a thin layer of detritus on frequency characteristics. a) Velocity profile (extra layer shaded); b) frequency characteristics (numbers on curves are focal depths in km).

earthquakes with a depth of 200 km and more, the observation of higher modes is very improbable in the part of their spectrum where the dispersion is mainly determined by the form of the crust; for periods $T < T_k'$, the intensity of Love waves is too low; they can only be generated in very strong earthquakes in which case they will be masked by S waves; for periods $T > T_k'$ the existence of Love waves is determined by the mantle.

Extra calculations also yield the following information:

4. The presence of a low-velocity layer with $h = 0$ leads to the appearance of an extra sharp-resonance peak in $\bar{V}_k(T)$ at periods shorter than 2 sec; this peak corresponds to Love waves in a sedimentary layer. When distance of the source below this layer increases, there is a sharp drop in the amplitude of the peak and it then completely disappears.

5. Frequency characteristics of models with surface layers and internal waveguides in the crust (types VIIIa, VIIIb, VIIIc) have some specific features (especially for $k \geq 3$) stated in general form in Paragraph 8, Section 3. For $h = 0$, between the short-period peak of $\bar{V}_k(T)$ ("waves in sediments") and the main pass band ("waves in the crust"), $\bar{V}_k(T)$ has sharp intermediate peaks and also intervals where it falls sharply. The intermediate peaks correspond to resonance phenomena in the waveguide-sedimentary layer system, the rapid drops correspond to a concentration of energy in the waveguide (channel waves in the crust). Thus for the third mode in models of type VIIIa we have the following results: the main pass band is in the range 3-8 sec; the intermediate peak of $\bar{V}_k(T)$ occurs for a period of about 1 sec; that corresponding to the sediment, at 0.5 sec. For higher modes there are intermediate peaks for periods of 1 sec and shorter. As the source becomes deeper the height of the intermediate peaks, relative to the shorter-period peaks due to the sedimentary layer, continuously increases.

6. The transformation of frequency characteristics in period ranges in which the waves do not penetrate the mantle to models with different crust thicknesses is made just as in the case of the fundamental mode. This also refers to models with different sedimentary layer thicknesses for periods at which the waves do not penetrate into the crust.

§5. Dispersion

In this paragraph, in the light of general considerations concerning the formulation of the inverse problem on pp. vii-viii, we consider the possibility of investigating the earth's crust using the dispersion of Love waves. In our investigation we use the experimental numerical method — the comparison of dispersion curves of the first three Love-wave modes calculated for various models of the continental crust.

It is sufficient to consider the eight crust models shown in Fig. 6A (with or without surface layers).

Detailed calculations show that, for surface waves, it is sufficient to consider the three models shown in Fig. 6B, while for the upper mantle it is sufficient to investigate the effect of the initial velocity b_M and the velocity gradient db/dz directly under the crust.

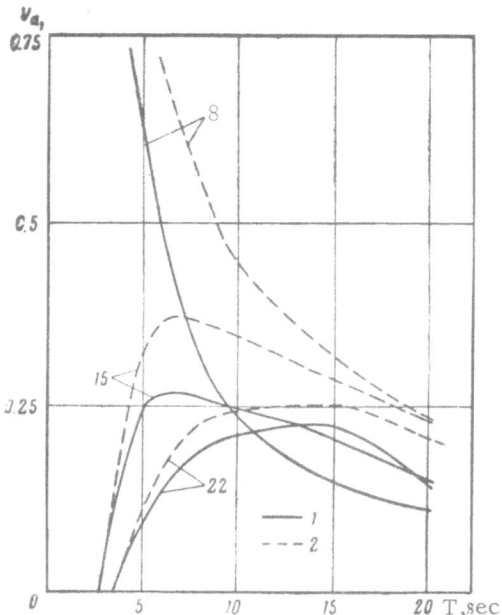

Fig. 10. Frequency characteristics $\bar{V}_{d_1}$ (T, h) for a dipole with a moment (model IIb). 1) E/G = 0.06; 2) E/G =0.5; numbers on curves are focal depths in km.

The total thickness H of the crust and the initial velocity b_0 in a "crystalline" crust were fixed for the calculations (H = 30 km, b_0 = 3.6 km/sec). The use of bilogarithmic coordinates, however, permits the extension of all conclusions to other models with changes of scale that agree with the scaling laws (Paragraph 4, Section 3.)*

1. The Fundamental Mode; Crusts without Surface Layers

Figure 12 shows dispersion curves of the phase velocity $v_1(T)$ and the group velocity $C_1(T)$ of the first mode in various models of a crust of thickness H = 30 km; the velocity distribution in the mantle is taken in all cases from the Jeffrey-Bullen results [37]. The following conclusions are obtained from a comparison of the curves:

1. The dispersion curves for different models are practically the same for very short and very long periods (T < 2 sec) and (T > 100 sec respectively). The first conclusion is due to the presence of the common asymptote v = b_0 for T → 0. With the exception of model VIII, the second is due to the fact that the structure of the mantle is identical for the different cases in Paragraph 2, Section 3. The fundamental mode begins to penetrate the mantle (for v > b_M) at periods of 50-60 sec and this penetration increases with further increases in T. It is easily calculated that, in the case of a thicker crust (H = 40 − 45 km), the structure of the crust can strongly influence the dispersion for T < 70 − 80 sec while it has practically no influence for T > 100 sec when the energy is mainly transmitted to the mantle.

2. At intermediate periods the difference between the models is as follows:

a) Models with mean velocities that only differ slightly (II, III, IV, VII, VIII) have similar dispersions, especially for T > 20-25 sec).   Only when the period is shorter than 10-15 sec can we attempt to distinguish the gradient models II and III from IV and VII by means of the weak group-velocity maximum. However extra calculations show that no specific intense vibrations can be associated with this maximum.

b) The model with the waveguide (VIII) differs appreciably in dispersion only when T < 15 sec, in which case the phase velocity is lower than that for other models (0.5-0.3 km·/sec) and the group velocity is 0.1-0.35 km/sec. Here however there is an extremely strong waveguide: the velocity in this waveguide (about 0.3 km/sec) is considerably lower than at the surface.

c) Models with a high gradient (V, VI) and with a homogeneous crust (I) differ strongly from the other models. This however is easily compensated for by a variation of the crust thickness or the transverse wave velocity in the crust. This assertion is confirmed by Fig. 13, in which bilogarithmic coordinates are used for the dispersion curves $v_1(T)$ and $C_1(T)$ for models I, V, and VII with the same velocities b_0 but with different thickness (25, 45, and 30 km respectively). We see that the difference between the dispersions in models I and VII or V and VII does not exceed 0.07 km/sec. Analogous results can be obtained for the variation of the

---

*Strictly speaking this leads to a proportional distortion of the mantle. To determine the undesirable errors introduced into the dispersion, a comparison was made of calculated results for models with distorted and undistorted mantles (with a fixed crust). It appeared that, for vibration periods at which the depth of penetration of the waves did not exceed H, the errors were not significant; they were smaller than 0.05 km/sec for variations in the depth scale by a factor of 1.5 and variations in the velocity scale by 5%.

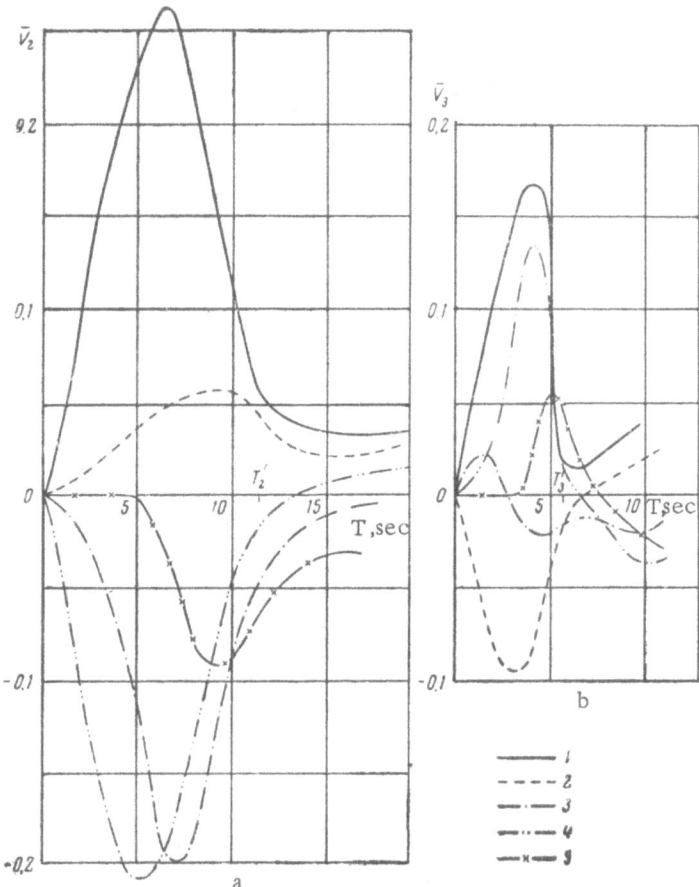

Fig. 11. Frequency characteristics $\bar{V}_K(T, h)$ for model I of the crust
and model 1 of the mantle for various focus depths. a) Second mode;
b) third mode; 1) h = 0 km, 2) 10 km, 3) 20 km, 4) 30 km,
5) 50 km.

velocity. Only model VI with an anomalously high gradient, improbable from many points of view, maintains
its difference for thickness and velocity variations in a reasonable range.

Additional calculations also show the following.

3. A variation of b_M [the transverse-wave velocity directly under the crust ($b_M = b(H+0)$] between
4.35 and 4.6 km/sec noticeably influences the dispersion in the crust at periods above 35-40 sec, while the
phase velocity varies from 0.05 to 0.07 km/sec and the group velocity varies from 0.10 to 0.12 km/sec. A
variation of b_M of the order of 0.1 km/sec is not realizable in practice. Variations of the velocity gradient
db/dz directly under the crust in the range $0.3-0.5 \cdot 10^{-2}$ sec^{-1} have less effect than the above variation of b_M;
it becomes realizable only for T > 50-60 sec.

4. It is impossible to choose one of the models under consideration and to determine simultaneously its thickness and initial (or mean) velocity. For example, we will successively consider each of the theoretical curves
for $v_1(T)$ or $C_1(T)$ (Fig. 12) for periods from 15 to 60-70 sec for various models, observed with an error of the
order ± 0.05 km/sec, and see the two-layer model VII with the equivalent dispersion. In this model we specify
as usual the velocity, density and thickness ratios and seek two parameters: the total thickness H_e, and the
surface velocity b_{0e} or the mean velocity $\bar{b}_e$. Comparing the resulting parameters with the "real" parameters,
i.e., with those corresponding to the models for which $v_1(T)$ or $C_1(T)$ have been calculated, we can estimate the

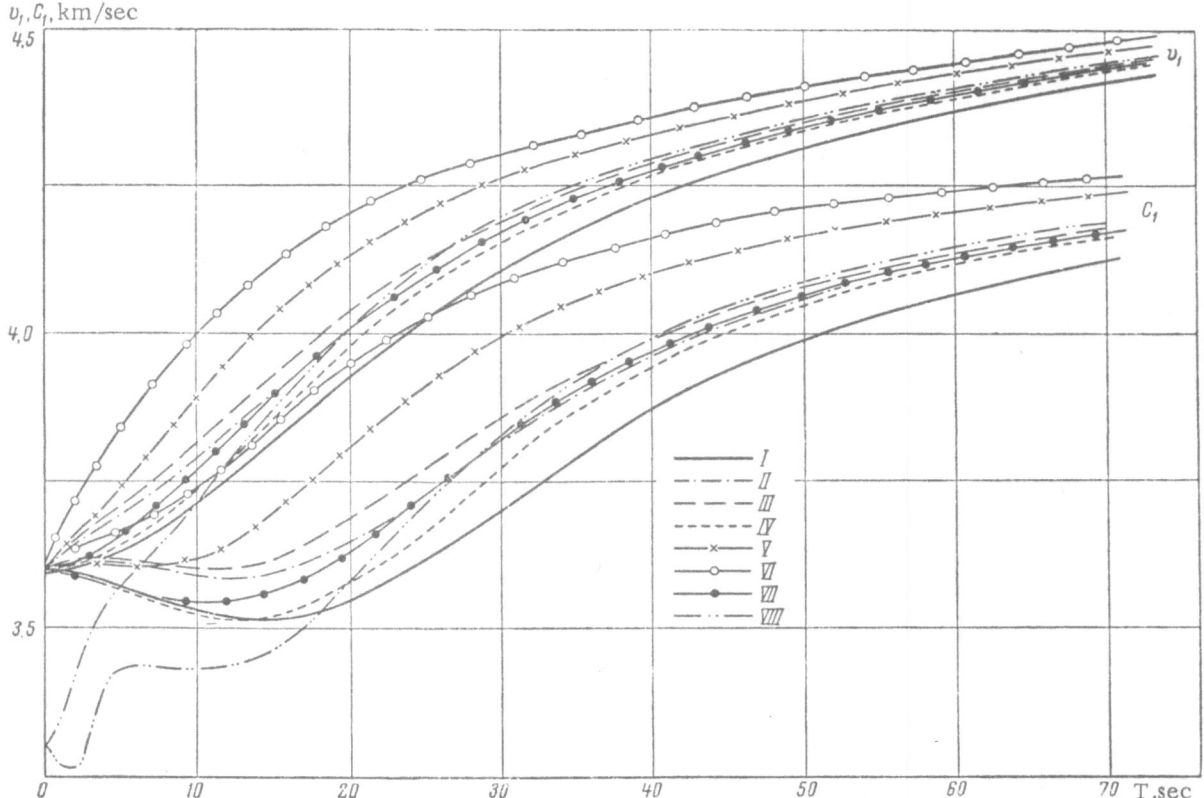

Fig. 12. Dispersion phases and group velocities for the first mode in models I-VIII of the earth's crust.

possible errors in the determination of these parameters due to the incorrectness of the choice of the model of the crust. The comparison is shown in Table 2. The models differ insignificantly in their mean velocities (by not more than 0.1 km/sec) but they differ appreciably but not too greatly in their total thickness (by 15-20%). Only for model VI, which has a very high velocity gradient, is it impossible to select the corresponding two-layer equivalent model.

The interpretation of dispersion curves is usually simpler: The velocities $\bar{b}_e$ and b_{0e} are fixed and only one parameter is sought — the thickness of the crust. We denote the crust thickness found in this way by H'_e. In Table 2 we give values of H'_e obtained for the fixed value $b_0 = 3.6$ km/sec which is the actual value. Here the errors in the crust thickness are naturally larger.

Hence, with the available accuracy of determination of phase and group velocities ($\pm$ 0.05 km/sec), we can estimate only the mean transverse-wave velocity in the crust with an error less than 5-7% and the total thickness of the crust with much lower accuracy — with an error of the order 15-25%. It is evidently impossible to obtain more detailed information concerning crust structure from the first mode. The phase-velocity curves have a lower resolving power.*

We now consider whether it is possible to raise the resolving power of the fundamental mode of Love waves by using a shorter section of the spectrum (T < 15 − 20 sec). For such periods the dispersion for different models differs more strongly, but it cannot be affected by surface layers with lower velocities.

---

*We note that phase-velocity curves have a certain advantage — they are determined on a very small base and so they depend less on the horizontal inhomogeneities of the medium.

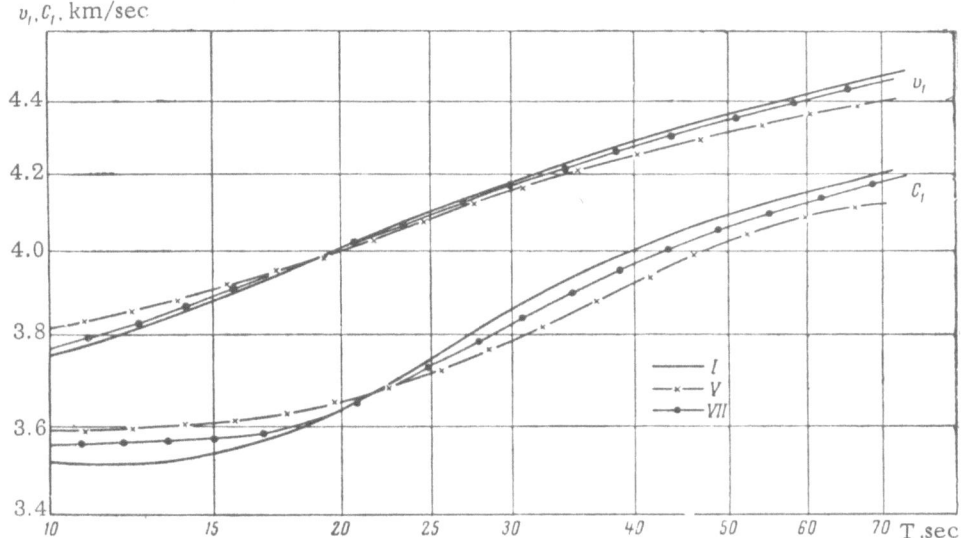

Fig. 13. Phase- and group-velocity dispersions of the first mode in model I (H = 25 km),
V (H = 45 km) and VII (H = 30 km).

Table 2. Parameters of Two-Layer Models of Type VIII Equivalent in Dispersion
to Models I-VIII (actual values for all models: b_0 = 3.6 km/sec, H = 30 km)

№	$\bar{b}$, km/sec *	$\bar{b}_e$, km/sec	b_{0e}	H_e	H_e'
I	3.60	3.65	3.56	34	36
II	3.70	3.71	3.62	30	31.5
III	3.73	3.73	3.64	30	31.5
IV	3.67	3.64	3.55	29	26
V	3.86	3.75	3.66	25	20
VI	3.96	no equivalent model			
VIII	3.73	3.63	3.54	28	30

*"Actual"

2. The Fundamental Mode; Crusts with Surface Layers

In Fig. 14 we compare the dispersion in models IIb (with a sedimentary layer with a gradient), IIc (with a surface layer), and II (without a weakened layer). We see that the sedimentary layer considerably changes the behavior of the dispersion curves for periods lower than 25-30 sec. The asymptote of the $v_1(T)$ curves in this case is the minimum velocity in the sedimentary layers; thus Love waves do not penetrate the crust for T < 7 - 8 sec (the phase velocity is lower than the transverse-wave velocity in a crystalline crust)(see pp.20-21). The $C_1(T)$ curves have a low minimum for T ≈ 3 − 5 sec. For long periods (up to 25-30 sec), for which penetration of the crust commences, the phase and group velocities increase with T considerably more rapidly than for a crust without a sedimentary layer.

A similar but less marked effect is caused by transitional layers. Hence the presence of weakened layers in the upper part of the crust levels out the differences in dispersion for different models of the crust at periods shorter than 15-20 sec. Only if the structure of these layers is well known can we attempt to distinguish these models by using the dispersion of the fundamental mode for the indicated periods.

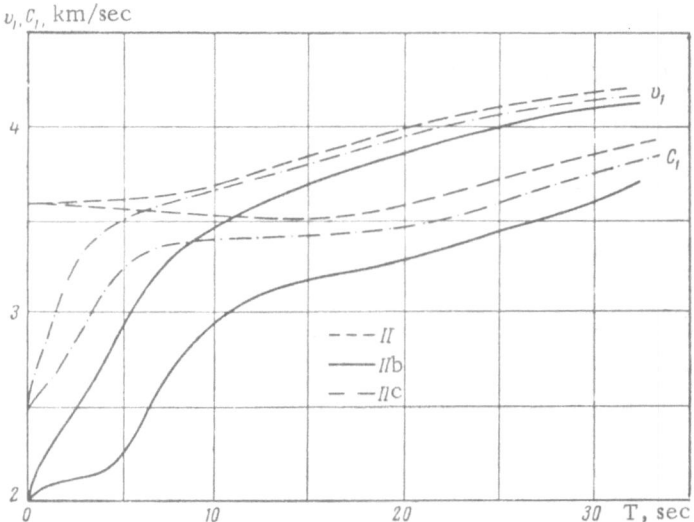

Fig. 14. Dispersion of the first mode in models with a surface layer (IIb, IIc) and without surface layers (II).

3. Higher Modes

The phase and group velocities for models I-VIII of a crystalline crust without any surface layers are shown in Fig. 15a, b (second mode) and Fig. 16a, b (third mode). The following differences between the first and higher modes are clearly indicated by these graphs:

1. The period interval in which the effect of the crust is felt is considerably narrowed, while the dispersion curves for the transition to propagation in the mantle are sharply bent. The position of this bend is the most reliable information from which the quantity H can be determined. Thus the second mode of Love waves is propagated in the crust for periods T sec $\leq (0.35-0.40)$ H km; for large T the waves penetrate the mantle, and the crust structure ceases to influence the dispersion.

In the third mode, the part of the spectrum generated by the influence of a crust of the same thickness is narrower (T sec ≤ 0.15 H km) and the bend in the $C_3(T)$ curve for the period T ≈ 0.15 H is still narrower.

2. The dispersion for different crust models shows a greater variation for different models than in the case of the fundamental mode, especially in the vicinity of minima of $C_k(T)$.

But the real record in the corresponding time intervals can be very complex due to the superposition of vibrations with only slightly differing periods but with different dispersions. Hence the usual method of analysis, which is to process this part of the record manually without any use of spectral analysis, can hardly be effective. The decomposition of a complex interference record into a set of modes and the determination of their amplitude and phase spectra requires the application of special methods of spectral and correlation analysis; such an analysis requires a numerical record of the surface waves.

3. Sedimentary or transitional layers have a strong influence on the dispersion of high modes only in a narrow range of short periods: up to 1.5-2 sec for the second mode and up to 0.5-1 sec for the third mode. This is illustrated in Fig. 17, in which a comparison is made of the dispersion of higher modes in models with and without a weakened layer.

In this region the curves for $v_k(T)$ and $C_k(T)$ fall sharply with decreasing T to the asymptotic value, i.e., to the velocity at the top of the sediments. Just as for the transition from the crust to the mantle, the transition from sedimentary layers to a crystalline crust is very clearly indicated (in the form of a sharp bend in the dispersion curves). The subsequent course of the curves for large T is independent of the presence of the sediments.

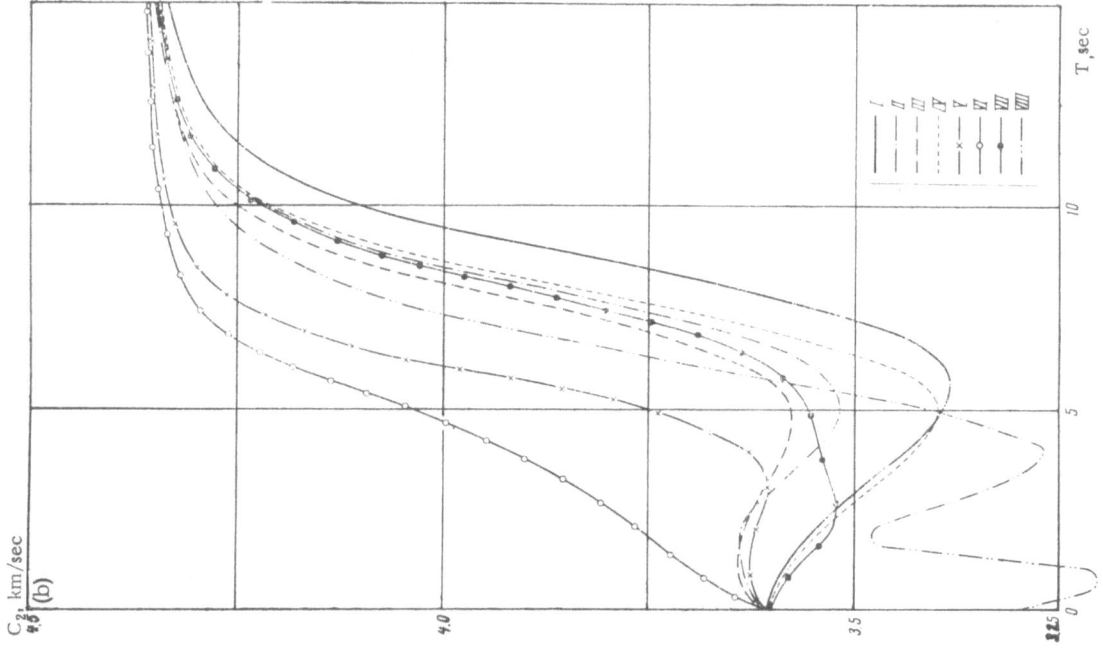

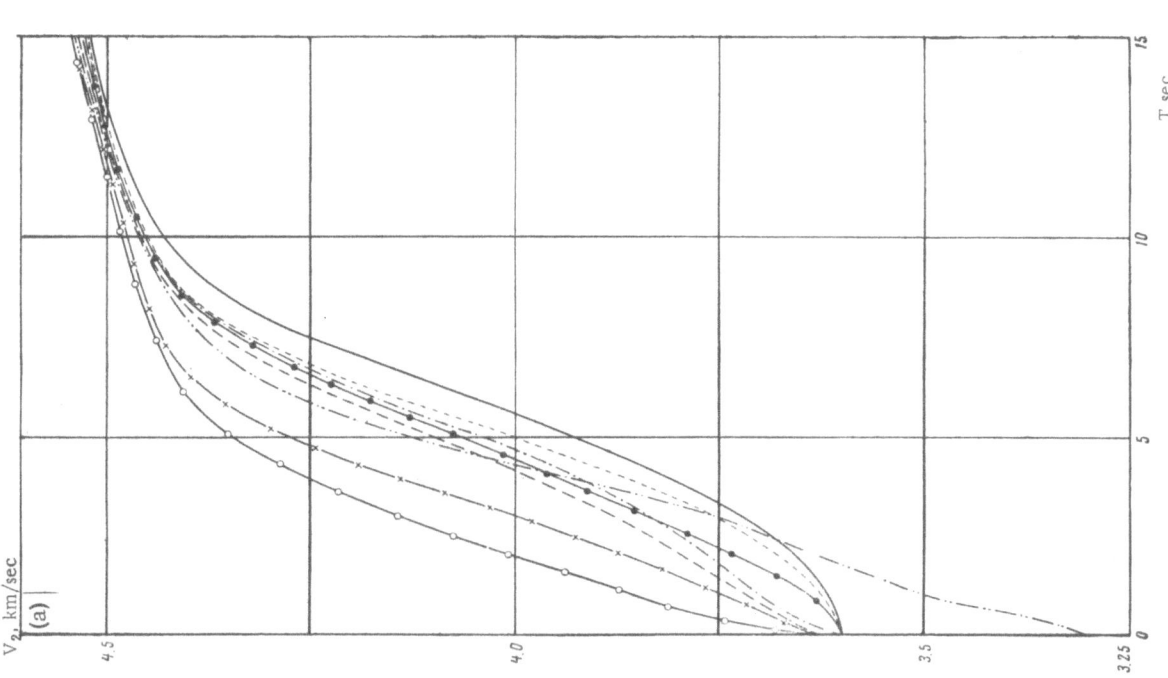

Fig. 15. Dispersions of the second mode in models I–VII of the crust. a) phase velocity; b) group velocity.

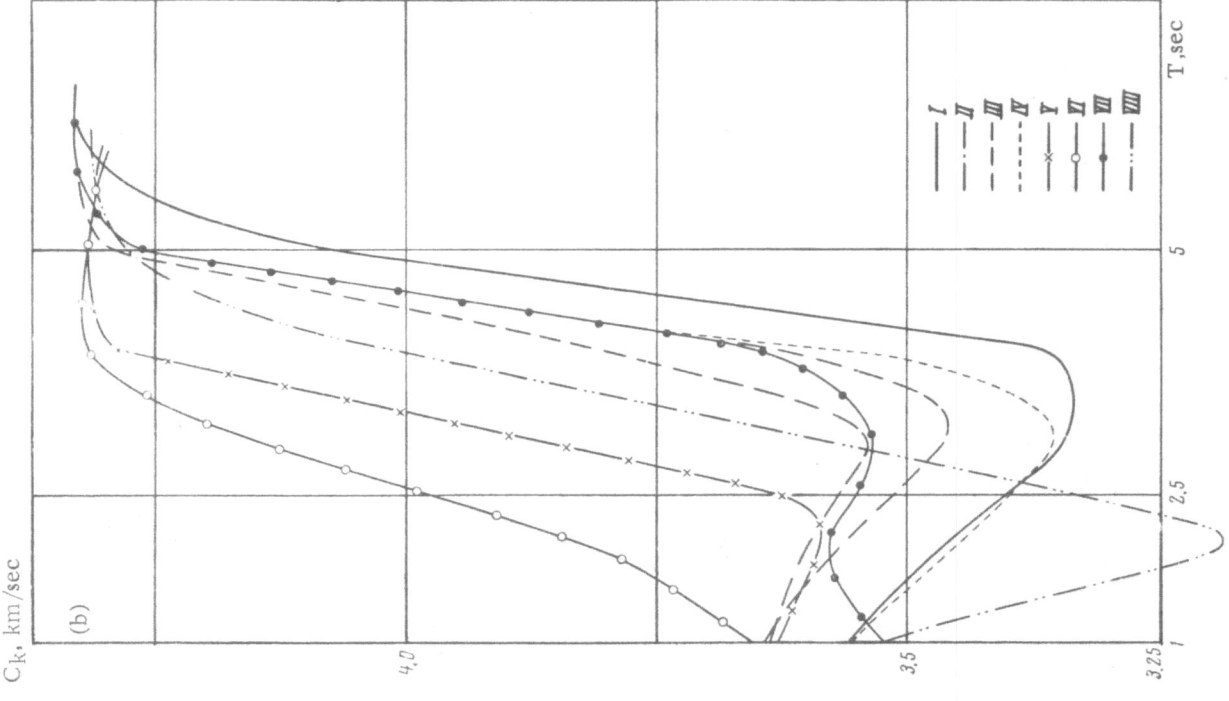

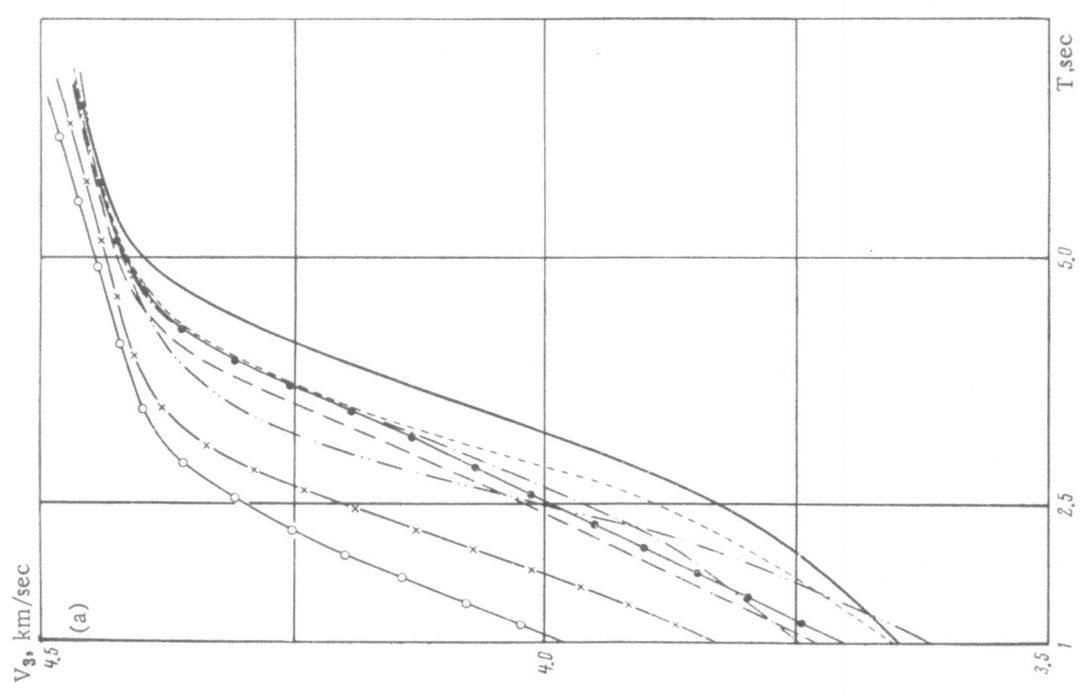

Fig. 16. Dispersions of the third mode in models I–VIII of the crust. a) phase velocity; b) group velocity.

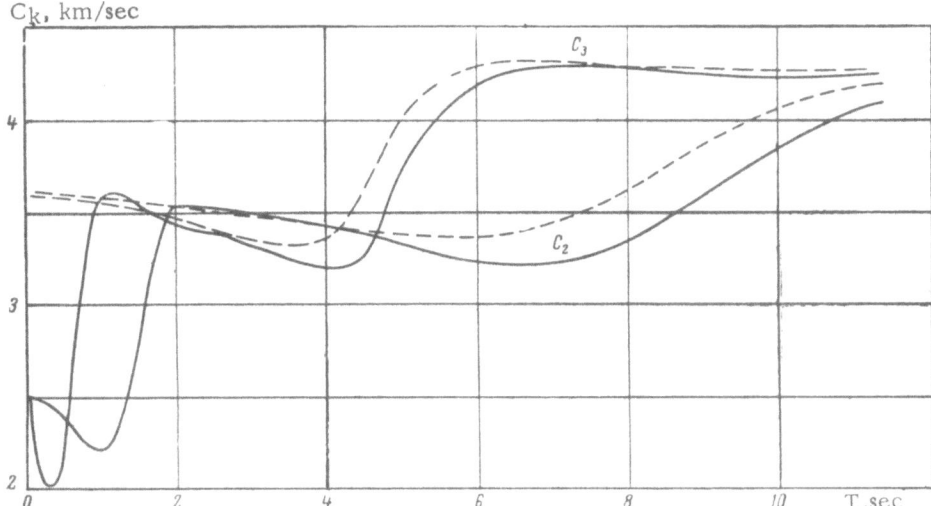

Fig. 17. Dispersions of group velocities of the second and third modes in models of the crust: continuous curves, with a 2km sedimentary layer; dotted curves, without a sedimentary layer.

Hence surface layers do not make the difference between the dispersion of higher modes for different crust models.

Extra calculations yield the following information:

4. An increase in the velocity b_M at the top of the mantle influences the steepness of the slope of the long-period branch of the curves $C_k(T)$ even more strongly, but it has only a weak effect on phase velocities. The gradient in the mantle has practically no effect on the dispersion to the left of the point of inflection.

5. The presence of an internal waveguide in a crust covered by sediment introduces a specific complication in the behavior of the dispersion curves of high modes (see Section 3, Paragraph 8). In the phase-velocity curves for periods shorter than 3-4 sec there are steep sections at the ends of which the curves for $v_k(T)$ and $v_{k+1}(T)$ approach one another. In the group-velocity curves, the regions where the phase-velocities approach one another correspond to sharp local minima of the group velocity $C_k(T)$. (For model VIIIa, for example, the longest-period minimum of this type occurs for the third mode at periods of 1.1-1.2 sec.) It is plain from the conclusions of Section 4 that resonance peaks of $\overline{V}_k(T)$ correspond to these minima; they describe the specific property of waves of long duration with a quasistationary period. However, because of the high absorption of such short periods, the possibility of recording these waves at distances of more than $10°$ is doubtful. At short distances their detection will be hindered by the presence of transverse and surface waves from other modes or other parts of the spectrum of the same mode. A similar but more easily detected phenomenon related to the presence of a waveguide in the mantle will be considered in more detail in Section 8.

§6. The Theoretical Wave Picture

The actual record obtained of Love waves is the result of observing several modes. The conditions of the superposition of these modes in a given time interval, even under the ideal conditions we have used here, are determined by a number of factors that can be classed in three groups:

1) the nature of the source—the focal depth, the directive properties of the radiation, and the space-time spectrum of the source;

2) conditions at the receiver—the epicentral distance and frequency characteristics of the recording apparatus;

3) the structure of the medium, i.e., its phase-and group-velocity dispersion and its frequency characteristics (see Section 3).

The information in this and the preceding sections can be used to take account of these factors. For definiteness we will assume that the amplitude spectrum of the source is identical at all frequencies and that the apparatus is equally sensitive to vibrations with $T < 50$ sec and strongly attenuates vibrations with longer periods.

We estimate the possibility of distinguishing the different modes in a period range corresponding to the propagation of Love waves in the crust. We use the names, "short-," "medium-," and "long-period" sections of each mode of vibration for periods that are considerably shorter than, approximately equal to, and considerably longer than the period of the main group-velocity minimum.

Crusts without Surface Layers

For definiteness, we assume that the structure of the crust is given by model II and the crust thickness is $H = 35$ km. It is clear from Fig. 18 that, for a given epicentric distance r, during the time interval from $t_1 = r/b_M$ to $t_2 = r/b_0$ there will be simultaneously recorded:

a) the long-period section of the fundamental mode; the period will decrease with the time of-arrival from 50 to 20-25 sec;

b) long-period sections of higher modes; the period decreases approximately from 10 to 8 sec for the second mode and approximately from 6 sec to 5 sec for the third. The relation between the intensities of the various modes varies in a complex fashion with the focus depth, but the fundamental mode is always more intense.

In this time interval the first mode can in many cases be distinguished visually, since its periods are much longer than the periods of higher modes. It is difficult to distinguish the other modes visually; filtering must be used to distinguish these modes.

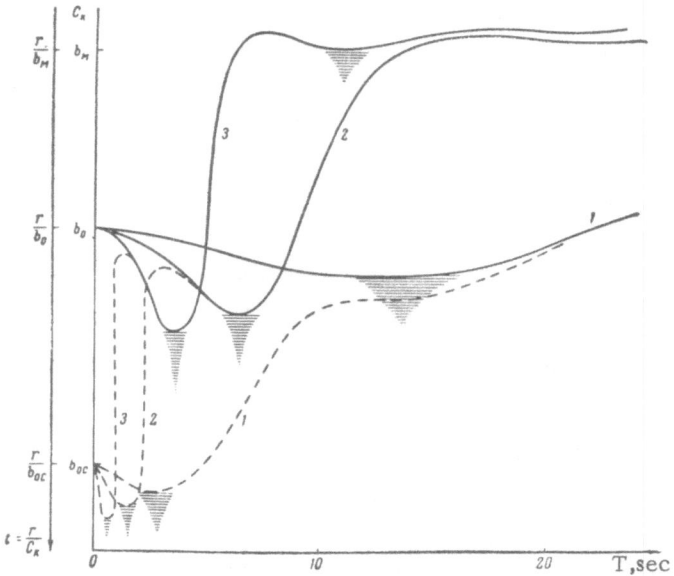

Fig. 18. Group-velocity dispersion of Love waves in the earth's crust (schematic). Continuous curves, a crust without a surface layer; dotted curves, with a surface layer; shaded regions correspond to Airy zones; numbers on curves are mode numbers.

In the time interval from t_2 to $t_3 = r/b_0 - (0.3 - 0.4$ km/sec) we can simultaneously record:

a) the short-period ($T < 5$ sec) and the medium-period sections of the fundamental mode including the Airy phase ($T = 12 - 15$ sec);

b) the analagous part of the higher modes with overlapping periods in the range $T < 7 - 8$ sec. Since they have intensities that are relatively close to one another for periods that differ only slightly, the higher modes here form a complex interference combination; the expansion into modes by simple methods (for example narrow-band filtering is hardly practicable.

The contribution of the separate modes to the over-all interference vibrations depends essentially on the focal depth (when the depth increases the contribution of the fundamental mode appreciably decreases), and also on the absorptive properties of the medium (selective absorption at sufficiently long distances from the source leads to a loss of short-period components).

In Fig. 19a, b we show theoretical seismograms $k = 1$ and $k = 2$ for $r = 6000$ km for various focal depths (40 and 75 km). The vertical arrows on the right divide each seismogram into two parts: the left-hand part was calculated by means of formulas (3.6) and (3.7) of the stationary-phase method, the right-hand part was calculated by using Airy functions and (3.8). In the junction region checks were made in separate cases by using the more accurate formulas from [30].

The seismograms show that the duration of vibrations in the Airy phases is approximately the same for $k = 1$ and $k = 2$, and for the given value of r this duration is approximately 20 sec. (Here and below the duration is the time interval during which the vibration amplitude exceeds one tenth of the maximum in the phase under consideration.)

Crusts with Surface Layers

The presence of surface layers leads to an extension of the record for $t > t_3$ and essentially changes the relation between the different modes for $t > t_2$ (see the dotted curves in Fig. 18).

In the interval t_2 to t_3 there will now be recorded:

a) the medium-period section of the fundamental mode; the period decreases with time from 20-25 sec to 10-12 sec;

b) an analogous part of the higher modes with overlapping periods in the range 2-7 sec. In this case the fundamental mode, as above, can be separated from the higher modes by using the longer periods.

In the interval $t > t_3$ for $h < H$ we can record:

a) short-period parts of the fundamental mode with $T < 10$ sec (they are dispersed in the sedimentary layer);

b) short-period parts of higher modes with periods $T < 2$ sec. The latter must be rapidly damped with increasing distance due to absorption, and so the fundamental mode must predominate for foci in the crust in this time interval. This is important in the application of the methods described in Section 10 for estimating the focal depth from the spectrum of the fundamental mode.

For large focal depths, all the waves in the time interval $t > t_3$ must be very weak.

Theoretical seismograms of the fundamental mode for a crust with a surface layer are shown in Fig. 31.

Summing up the results of Sections 5 and 6, we reach the following conclusions:

1. The mean velocity of transverse waves in the crust can be determined with an error of less than 5-7% and the crust thickness can be determined with an error of about 15-20% from the first mode of Love waves if the phase and group velocities are determined with an error of the order of 0.05 km/sec in the period range 20-70 sec.

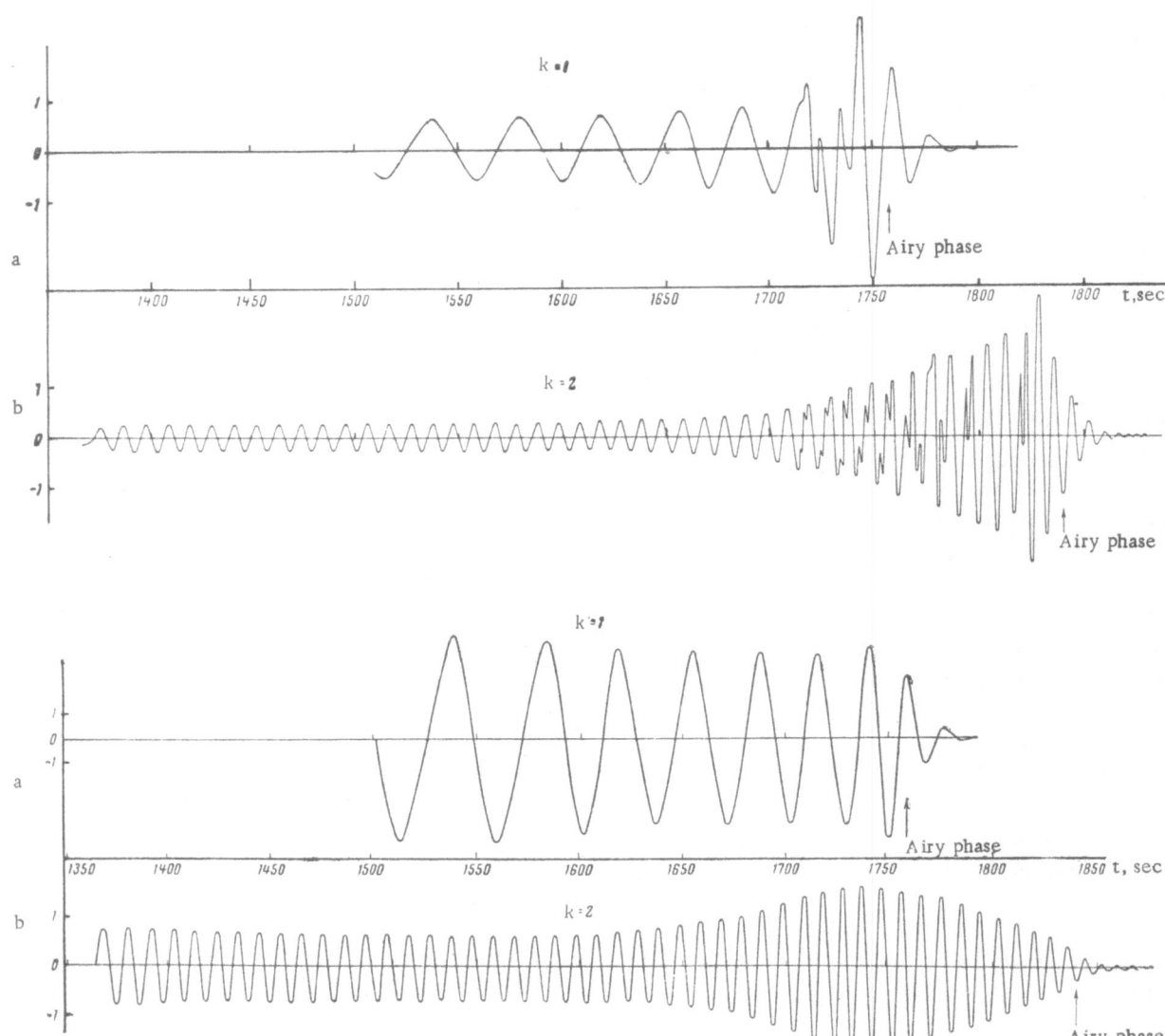

Fig. 19. Theoretical seismograms of the first and second mode of Love waves for Model II of the crust (H = 35 km, b_0 = 3.5 km/sec) for the mantle 2 with epicentric distance 4 = 6000 km. The source is a concentrated force with a uniform spectrum. a) focal depth 40 km; b) focal depth 75 km.

2. If the structure of the surface layers is known, these layers can be taken into account in the calculations and some characteristic features of the structure of the crystalline crust can be determined from the dispersion of fundamental mode in the period range 15-30 sec (the presence of strong waveguides or of zones of anomalously high gradient close to the surface, etc.).

3. To obtain more accurate information concerning the structure of the crust we can use the dispersion of the second and third mode. Conditions favorable for such a calculation are the following:

a) the presence of surface layers ensuring the separation of fundamental and higher modes even when the periods only differ slightly;

b) a focal depth of the order of 30-40 km (this ensures a larger relative intensity of higher modes as compared with the fundamental mode for these periods than would be obtained from foci close to the surface). For crust thicknesses of 30-40 km, the recording of the second harmonic of Love waves in the crust is possible in the range $T < 12 - 13$ sec, and the recording of the third is possible in the range $T < 7 - 8$ sec.

4. A crust with a waveguide is characterized by a considerable drop in the phase and group velocities of the fundamental mode for periods shorter than 15 sec. Vibrations with periods shorter than 2 sec are concentrated in the waveguide and do not reach the surface. The presence of a surface layer masks the effect of a waveguide on the dispersion of the fundamental mode, but leads to characteristic features in the higher modes at periods shorter than 2 sec. In particular, specific vibrations with quasistationary periods and long duration can be generated corresponding to resonance in the system consisting of a surface layer and a waveguide in the crust. The detection of such waves in real seismograms, however, is hardly practicable because of the strong absorption of short-period vibrations in the crust and the presence of interfering waves (volume waves and other modes of the surface waves).

CHAPTER III

LOVE WAVES IN THE UPPER MANTLE

In this chapter we investigate Love waves with periods for which they actively penetrate the upper part of the earth's mantle.

It follows from Section 3 (Fig. 2) that these periods must be sufficiently long for the phase velocity of Love waves to be larger than the velocity of surface waves in the part of the mantle under consideration. It is, however, not satisfactory to use periods that are too long: As the period increases the structure of the mantle begins "to average itself out." It is clear that the maximum attainable period decreases in length with increasing mode number.

We confine ourselves to the consideration of the range $v_k < 5.00$ km/sec, corresponding to penetration of Love waves to 400 km or 0.06 times the radius of the earth; this condition is satisfied for periods shorter than 250 sec for the first mode, shorter than 50 sec for the second mode, and shorter than 25 sec for the third mode.

In Section 7 we consider frequency characteristics of the medium for various models of the upper mantle. In Section 8 the relation between possible fundamental properties of a profile (the presence or absence of zones of lower velocity, higher or lower gradient) and the features of various modes are investigated. Examples are given of the determination of a set of velocity profiles satisfying observations of S waves and Love waves. In Section 9 possible conditions for the recording of various modes are considered and examples of theoretical seismograms are given.

§7. Frequency Characteristics of the Medium

1. The Fundamental Mode

Figure 20 shows frequency characteristics $\overline{V}_1(T, h)$ for model I of a homogeneous continental crust (H = 30 km b = 3.6 km/sec) on two different models of the mantle — the Jeffreys-Bullen model (1) and the Gutenberg model (2) (Fig. 6C). Calculations were carried out for a simple-force type of source.

At periods longer than 60 sec, for which Love waves penetrate the mantle, $\overline{V}_1(T, h)$ increases monotonically with T. As the focus depth h increases $\overline{V}_1(T, h)$ decreases, and its rate of increase with T becomes more rapid. The maximum of $\overline{V}_1(T, h)$ for h = 0 is about 0.6 times its maximum (resonance) value for the "crust" part of the spectrum. Differences between the graphs for different models are not very great and are related principally not to the presence or absence of a layer with a lower velocity, but to the different velocities of transverse waves in the top of the mantle (4.35 km/sec for model 1 and 4.55-4.61 km/sec for model 2). The differences decrease as the focus depth increases.

In Fig. 20 frequency characteristics $\overline{V}_1(T, h)$ are shown for the model consisting of an oceanic crust (I) (H = 5 km, $b_0 = 3.55$ km/sec) on a Gutenberg mantle (2).

Here the Love waves penetrate the mantle for periods of 8-9 sec. For $9 < T < 15$ sec vibrations are mainly concentrated in the weakened layer and their intensity is low at the surface even for the optimal location of the focus — inside the layer with lowered velocity. Calculations show that in this range of periods $\overline{V}_1(T, h)$ does not decrease monotonically as h increases. When h increases $\overline{V}_1(T, h)$ first increases and then,

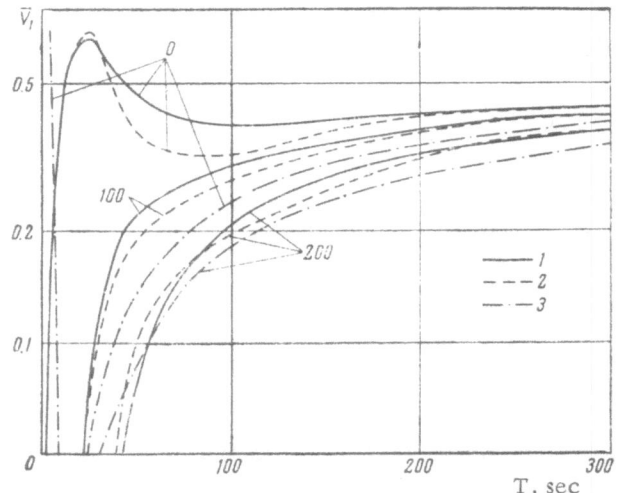

Fig. 20. Frequency characteristics $\overline{V}_1(T, h)$ of the medium
for different models. 1) continental crust I (H = 30 km) and
Jeffreys-Bullen mantle (Fig. 6C, 1); 2) continental crust I
(H = 30 km) and Gutenberg mantle (Fig. 6C, 2) oceanic
crust I (H = 5 km) and Gutenberg mantle; numbers on curves
give focal depths (km).

after reaching its maximum close to the axis of the weakened layer, begins to decrease. When T increases $\overline{V}_1(T, h)$ increases rapidly and max $\overline{V}_1(T, h)$ is displaced towards the surface and reaches it for T > 70 sec.

The difference between the graphs of $\overline{V}_1(T, h)$ for oceanic and continental models becomes insignificant only for periods in the range 200-300 sec.

2. Higher Modes

Figures 21a and 21b show frequency characteristics $\overline{V}_2(T, h)$ and $\overline{V}_3(T, h)$ of the medium for those models illustrated in Fig. 20 with h = 0 (focus at the surface) for the second and third modes.

We first consider models with a continental crust (curves 1 and 2). We see that the spectra of the higher modes decompose into two parts separated by a zone of low intensity. The left part corresponds to the crust, the right to the mantle. For a mantle without a waveguide (1), the values of $\overline{V}_k(T, 0)$ at the maxima and minima are commensurable, and an almost monotonic increase in intensity with increasing T is characteristic of waves in the mantle. In the case of a mantle with a waveguide (2), the values of $\overline{V}_k(T, 0)$ in the minimum zone are very small (a few orders smaller than in the zone of the left maximum); this corresponds to a concentration of energy in the waveguide. In this case in the third mode (Fig. 21b) there is, in addition to the main left maximum (waves in the crust), a further local maximum of $\overline{V}_3(T, 0)$ for larger T, corresponding to waves in the mantle. In the period range 8-10 sec, it has a sharply resonant character, and is separated from the crust part of the spectrum (T < 5 sec) by an interval of sharply lower values of $\overline{V}_3(T, 0)$.

In oceanic models with a waveguide (curves 3), the second mode also has a local maximum in the range 8-10 sec; the third mode has somewhat similar maxima. Only these maxima appear in Fig. 21, since the remaining part of $\overline{V}_k(T, 0)$ corresponds to incommensurably small values of $\overline{V}_k(T, 0)$. Calculations show that a decrease in the crust thickness H causes the maxima to move to the left, increases their amplitude, and decreases their width. These features of the behavior of $\overline{V}_k(T, 0)$ in models with a waveguide are indicated in the framework of the general description in Section 3 Chapter I; they will be investigated in more detail later.

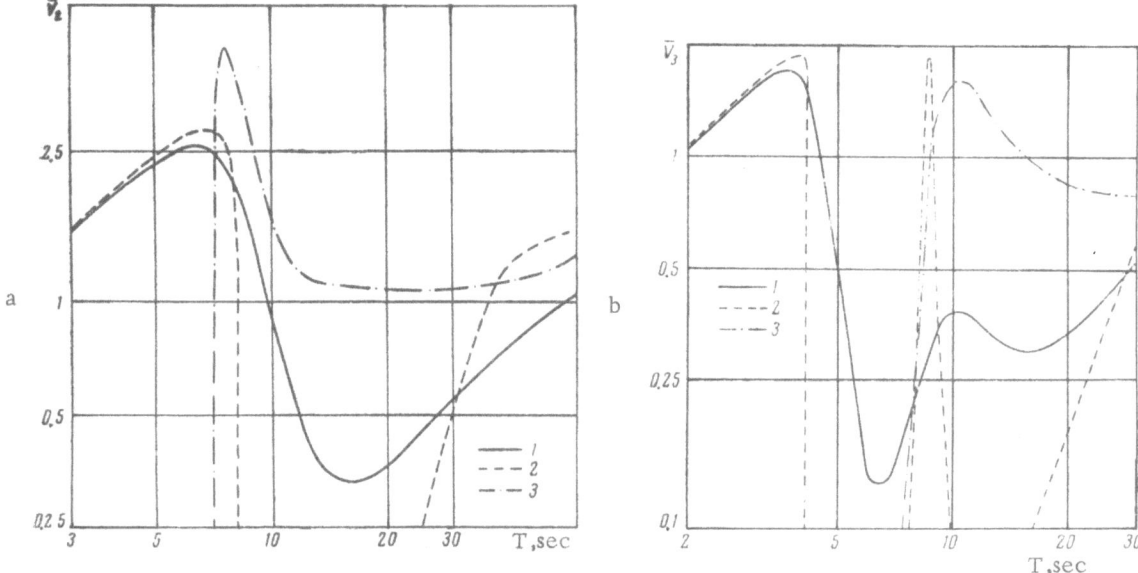

Fig. 21. Frequency characteristics $\overline{V}_k(T, 0)$ of the medium for the models of Fig. 20; focus at the surface. a) second mode; b) third mode.

When the focal depth h increases the shape of the frequency characteristics becomes more complex and regions of small and zero values of $\overline{V}_k(T, h)$ appear, the position of which depends on h. When h > H, the short-period section of the spectra corresponding to the crust gradually becomes much weaker than the long-period section describing waves in the mantle.

§8. Dispersion

In this section we investigate the possibility of using Love waves to determine the velocity profile in the mantle. As in Section 5, the investigation is carried out by the numerical experimental method — a comparison of dispersion curves for the first three modes calculated for various models of the mantle [44].

For our basic models we chose 22 profiles of the mantle from the 110 presented in [28] satisfying the travel-time curves for P waves for Europe [two profiles for each of the 11 types of longitudinal-wave velocity distributions a(z)]. The conversion to transverse-wave velocities was performed by using Gutenberg's results [8] and also by using the relation b = $a/\sqrt{3}$. The variation of the density with depth was in all cases taken to be that of Bullen's model A([37], Fig. 6B). Modell II of the earth's crust (Fig. 6A) was used, in which the thickness was 35 km. In this model the velocity b(z) increases linearly from 3.5 to 3.6 km/sec and the density $\rho(z)$ increases from 2.7 to 3.0 g/cm^3.

The profiles considered correspond to various types of mantle structure: with a waveguide directly under the crust; with a waveguide at a greater depth; without a waveguide but with various velocity gradients. Some examples of these profiles are shown in Fig. 22a and 22b.

According to Gutenberg's results, b/a is much smaller than $1/\sqrt{3}$ in the upper 200-300 km of the mantle, and this explains the main differences between the profiles in Fig. 22a and 22b.

1. The Fundamental Mode

Figure 23 shows dispersion curves of the fundamental mode in the 70-300 sec period range corresponding to the models in Fig. 22, and also some experimental results from [33, 35, 40, 38, 50]; these results unfortunately do not only refer to Europe, but were obtained mainly from observations of waves on very widely dispersed, mainly continental paths.

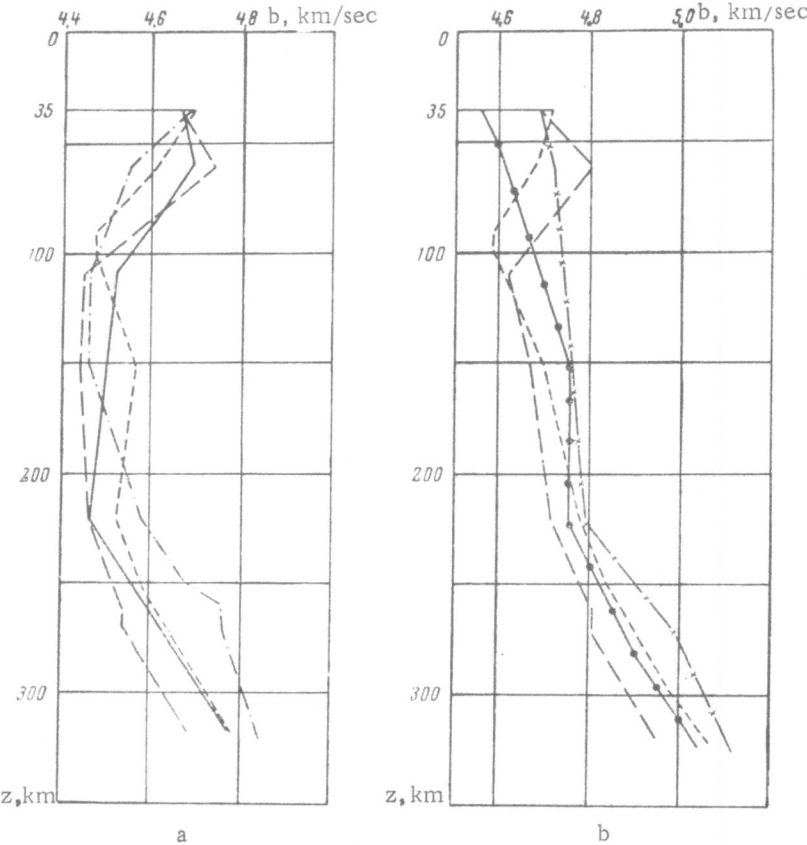

a   b

Fig. 22. Some profiles of the mantle satisfying the travel-time curve for
P waves [28]. Conversion of transverse-wave velocity. a) From Gutenberg's
results; b) with the assumption that b = $a/\sqrt{3}$.

A slow increase of the phase velocity and a very small change in group velocity occurs when T increases. For profiles with the gradient of the velocity b(z) close to zero or negative at some interval of depth below the crust, the group velocity is practically stationary for the period range 100 < T < 200 sec.

A comparison of experimental and calculated phase velocities yields only one concrete conclusion: Transverse-wave velocities in the depth interval 100-300 km decrease for all profiles in Fig. 22a, and increase for all profiles in Fig. 22b. In the profile which satisfies the observed dispersion, the velocity b(z) at depths from 100 to 300 km must be of the order of 4.6–4.8 km/sec, i.e., higher than that shown in Fig. 22, and higher than that given in the Gutenberg profile.

A more important method of obtaining information is to compare calculated dispersion curves. Figure 23 shows that the differences in dispersion for different profiles is not larger than the scattering of the experimental points. The maximum differences are for periods of from 60 to 100 sec. It is difficult to use these periods, however, since they are very sensitive to variations in the structure of the crust and the top parts of the mantle in extended traces. Separate characteristic features of the velocity profile are not sharply indicated in the dispersion curves. Hence all profiles have practically the same dispersion for the first mode. On the other hand these profiles correspond to very similar mean velocities

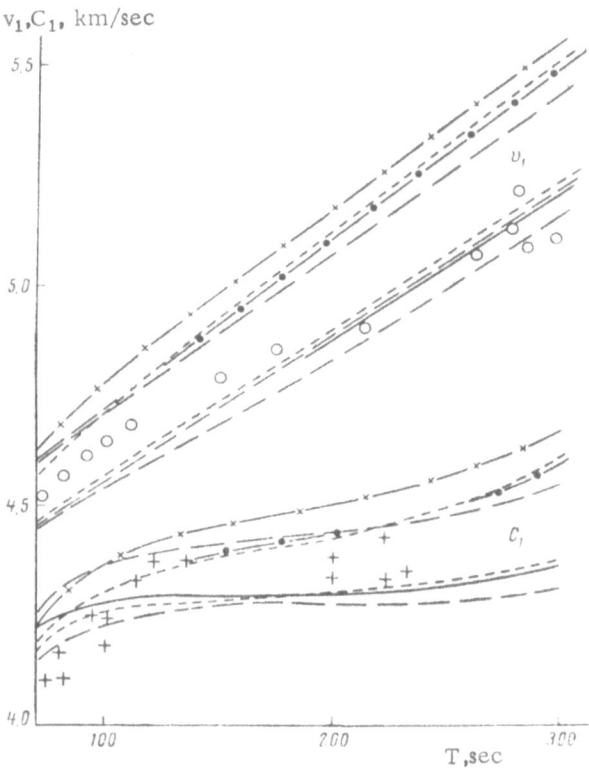

Fig. 23. Dispersion of the first mode in the following models of the
mantle. a) Velocity profiles of Fig. 22a; b) velocity profiles of
Fig. 22b; circles and crosses indicate phase and group velocities
respectively, obtained from experimental results.

$$\frac{\int_{H}^{z} b(z)\,dz}{z-H} \approx \frac{z-H}{\int_{H}^{z} dz/b(z)} \approx 4.75-4.80 \ \text{km/sec}$$

when $z \approx 300$ km, i.e., at a sufficiently large depth interval.

Thus dispersion of the first mode is determined by certain integral effects of the mean profile velocity.

2. The Inverse Problem

The observed dispersion curve of Love waves corresponds to a group of possible velocity profiles in the upper mantle. According to [28] the same is true of the observed travel-time curve for body waves. Since these groups are obtained independently, their intersection, i.e., the set of profiles satisfying both the travel-time curve and the dispersion, can contain fewer members than either of the two groups independently. We will use a specific example to illustrate a possible approach to the simultaneous interpretation of body and surface waves. This approach is described in detail in [28, 32]. It consists of using a mechanical device to map variants of profiles and to select sets of these profiles of the medium under investigation for which the theoretically calculated results (in our case the travel-time and dispersion curves) are sufficiently close to the observed results.

Original Observations. The original observations are the dispersion of the phase velocity (from [36]) and of the group velocity (from [40]) of the first mode of Love waves of the Nevada-New York path (Table 3) and the travel-time of S waves [46] for the north-eastern USA (Table 4).

The use of this travel-time requires some care. First, it was obtained as the mean of four earthquakes with different (and definitely positive) focal depths (40, 60, and 125 km); it is true that the points for different depths are well indicated on one of the travel-times. Second, there is a discontinuity with a sudden increase in the time of passage for the epicentric distance $r = 14°$. It was assumed that this discontinuity corresponded to a loop, but it is difficult to establish reliably which branch of a loop was recorded for $r > 14°$.

Profile Parameters. The profile of the crust and the upper mantle is obtained from values of $b(z)$ at depths $z_i (i = 0, ..., 8)$. Between these depths the profile is approximated for body waves by the Bullen law [38]

$$b(z) = b_i \left(\frac{R_0 - z_i}{R_0 - z} \right)^{k_i + 1} \tag{8.1}$$

(where $z > z_1$, R_0 is the earth's radius and k_i is an arbitrary constant), and for surface waves by two straight-line segments passing through the point $z = (z_i + z_{i+1})/2$, where they satisfy (8.1). The difference in the method of interpolation is of no importance in our case. The following limitations were imposed on the values of $z_i(km)$, $b(z_i)(km/sec)$, and $\rho(z_i)(g/cm^3)$:

Surface	$z_0 = 0$	$b(z_0) = 2.50$	$\rho(z_0) = 2.50$	
Base of sediments	$z_1 = 3$	$b(z_1 - 0) = 2.52$ $b(z_1 + 0) = 3.50$	$\rho(z_1 - 0) = 2.50$ $\rho(z_1 + 0) = 2.70$	
Base of crust	$37 < z_2 < 42$	$3.63 < b(z_2 - 0) < 3.87$ $4.55 < b(z_2 + 0) < 4.72$	$\rho(z_2 - 0) = 2.85$ $\rho(z_2 + 0) = 3.32$	
Waveguide possible	$50 < z_3 < 70$	$b(z_2 + 0) - 0.1 < b(z_3) <$ $< b(z_2 + 0) + 0.1$	$\rho(z_3) = 3.35$	(8.2)
	$80 < z_4 < 100$ $130 < z_5 < 150$	$4.35 \leqslant b(z_4) \leqslant 4.75$ $4.35 \leqslant b(z_5) \leqslant 4.75$	$\rho(z_4) = 3.37$ $\rho(z_5) = 3.42$	
Positive velocity gradient	$z_6 = 200$ $z_7 = 300$ $z_8 = 400$ $z_9 = 500$ $z_{10} = 600$	$b(z_5) \leqslant b(z_6) \leqslant 4.80$ $b(z_6) \leqslant b(z_7) \leqslant 4.90$ $b(z_7) \leqslant b(z_8) \leqslant 5.00$ $b(z_9) = 5.30$ $b(z_{10}) = 5.60$	$\rho(z_6) = 3.47$ $\rho(z_7) = 3.55$ $\rho(z_8) = 3.63$ $\rho(z_9) = 3.89$ $\rho(z_{10}) = 4.13$	

Hence to each profile there corresponds a point in the 12-dimensional space of the unknown parameters $b(z_2 - 0)$; $b(z_2 + 0)$; $b(z_i)$ $(i = 3 - 8)$; z_i $(i = 2 - 5)$.

Table 3

T_j, sec	$\overline{v_1}$, km/sec	c_1, km/sec	T_j, sec	$\overline{v_1}$, km/sec	c_1, km/sec	T_j, sec	$\overline{v_1}$, km/sec	c_1, km/sec	T_j, sec	$\overline{v_1}$, km/sec	c_1, km/sec
10	3.53	3.16	45	4.28	3.72	80	—	4.19	115	—	4.34
15	3.68	3.28	50	4.34	3.84	85	—	4.22	120	—	4.35
20	3.72	3.34	55	4.38	3.91	90	—	4.25	125	—	4.36
25	3.83	3.44	60	4.42	3.98	95	—	4.27	130	—	4.38
30	4.04	3.54	65	4.44	4.04	100	—	4.29	135	—	4.39
35	4.13	3.62	70	4.46	4.09	105	—	4.31	140	—	4.40
40	4.22	3.70	75	—	4.14	110	—	4.32			

Table 4

r_i^0	t,sec	r_i^0	t,sec	r_i^0	t, sec	r_i^0	t,sec
1	40	7	180	13	320	19	480
2	62	8	200	14	345	20	496
3	84	9	224	15	370	21	514
4	107	10	248	16	410	22	536
5	130	11	272	17	430	23	560
6	156	12	296	18	454	—	—

Formulation of the Problem. Our problem is to find, in the region of the space of unknown parameters defined by the inequalities (8.2), the subregion in which

$$\frac{1}{n_t - 1} \sum_i^{n_t} \left[\frac{t_i(r_i) - \bar{t}(r_i)}{\sigma[\bar{t}(r_i)]} \right]^2 \leqslant \bar{l}_1,$$

$$\frac{1}{n_v - 1} \sum_{\pm j}^{n_v} \left[\frac{v_1(T_j) - \bar{v}_1(T_j)}{\sigma[\bar{v}_1(T_j)]} \right]^2 \leqslant \bar{l}_2, \qquad (8.3)$$

$$\frac{1}{n_c - 1} \sum_j^{n_c} \left[\frac{C_1(T_j) - \bar{C}_1(T_j)}{\sigma[\bar{C}_1(T_j)]} \right]^2 \leqslant \bar{l}_3.$$

$$\max_i |t(r_i) - \bar{t}(r_i)| \leqslant \bar{l}_4,$$

$$\max_j |v_1(T) - \bar{v}_1(T_j)| \leqslant \bar{l}_5, \qquad (8.4)$$

$$\max_j |C_1(T_j) - \bar{C}_1(T_j)| \leqslant \bar{l}_6.$$

Here t, v_1, and C_1 are respectively the time of passage of S waves and the phase and group velocities of the first Love wave mode calculated theoretically; $\bar{t}$, $\bar{v}_1$, and $\bar{C}_1$ are values obtained from observations at the given number of points (n_t, n_v, n_c); r_i and T_j are the values of r and T for which t_1, v_1, and C_1 are given in Tables 3 and 4; $\sigma[x]$ is the mean-square value of x (the scattering of the experimental points); $\bar{l}_1 \ldots \bar{l}_6$ are the functions given in the table (giving the maximum permissible difference between calculated and observed values).

Method of Solution. The method is the same as that described in [28, 32]. In the region defined by the inequalities (8.2) we successively introduce pseudorandom points. For the profile corresponding to each point, theoretical values $v_1(T_j)$ and $C_1(T_j)$ are calculated by using the program described in Paragraph 9, Section 3, and values of $t(r_k)$ are calculated by using a program written by I. Ya. Azbel' for three different focal depths h (40, 70, 125 km). The conditions (8.3) and (8.4) are now checked. Separate profiles are sought that satisfy the observations of body waves (for at least one of the focal depths) and that satisfy the observations of surface waves. I. Ya. Azbel' wrote the program for the selection and comparison of body waves and V. P. Valyus wrote the program for surface waves. The following quantities were used in the calculations: $\sigma[\bar{v}] = \sigma[\bar{C}] = \bar{l}_5 = \bar{l}_6 = 0.1$ km/sec; $\sigma[t(r)] = \bar{l}_4 = 4$ sec for $r \leqslant 14°$ and $\sigma[\bar{t}(r)] = \bar{l}_4 = 20$ sec for $r > 14°$. The large value of $\bar{l}_4$ for $r > 14°$ is compensated for by the above indicated indefiniteness of the S-wave travel-times however, in spite of the large value of $\bar{l}_4$, far from all the profiles sorted satisfy the travel-times for $|r > 14°$.

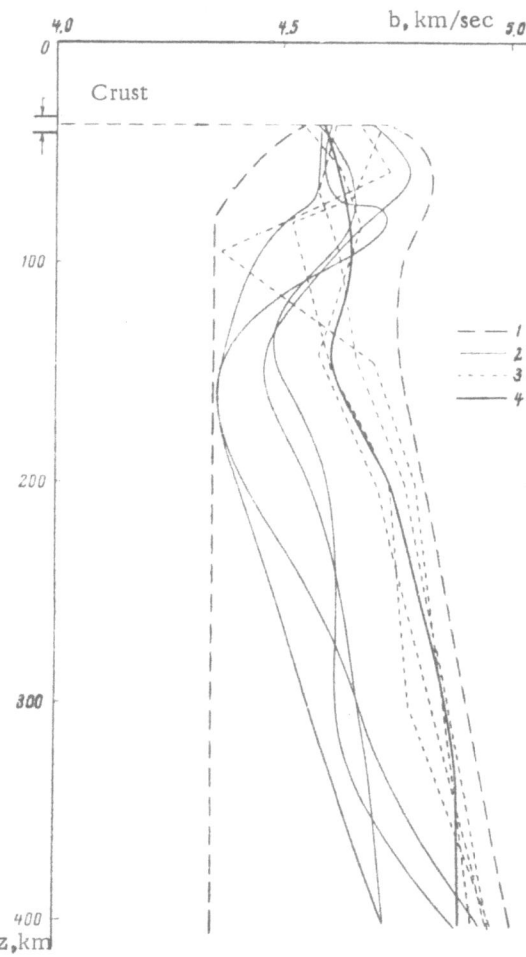

Fig. 24. The inverse problem for the upper mantle. 1) Contour of the region in which velocity profiles were sought; 2) profiles obtained from S waves; 3) profiles obtained from Love waves; 4) a profile satisfying both S waves and Love waves.

For these values of σ the selection of $\bar{l}_1$, $\bar{l}_2$, and $\bar{l}_3$ is not essential; profiles satisfying (8.4) also satisfy (8.3) for the minimal values $\bar{l}_1 = \bar{l}_2 = \bar{l}_3 = 1$.

Results Obtained. The results are shown graphically in Fig. 24. Four hundred profiles were selected. Twenty one of them satisfy the condition for body waves (all for h = 120 km) and eighteen satisfy the condition for surface waves. The sets of profiles satisfying body and surface waves not only do not intersect, but they hardly touch one another. We succeeded in finding only one common profile (the heavy curve in Fig. 24).

It would be premature, however, to assume that we have found the unique solution of the inverse problem. We must first investigate the effect of the chosen parametrization on the result. Will more solutions be found with a more flexible parametrization? We have evidently varied the structure of the crust excessively and this could lead to an unnecessary rejection of some mantle profiles. It should finally be noted that the travel-times and the dispersion curves were obtained in somewhat different regions, although it is true that there is no foundation for explaining a difference between corresponding profiles by a horizontal variation of the mantle — more evidence must be obtained for such a far-reaching assertion.

We describe this material only to illustrate our approach to the simultaneous processing of surface and body waves. In this connection it is interesting to note the following: The values of $b(z)$ in the interval $150 < z < 300$ km for most body-wave profiles are smaller than those for surface-wave profiles (this is in agreement with the material of Paragraph 1 of this section). It appears that at smaller depths the sets of profiles include any types of variants — with and without waveguides. However, the mean velocity for all profiles of body waves varies considerably less than the velocities in separate intervals of the depth. (For a more reliable travel-time of P waves, the mean velocities are practically identical for all profiles.) In a profile satisfying the conditions for both Love waves and S-waves, $b(z)$ must take its limiting value for $z > 150$ km; but at smaller depths there must be either a sharply delineated waveguide or an extension of the zone of relatively small values of $b(z)$. With increasing accuracy of experimental results, this approach could be more useful and could have more resolving power in detecting waveguides than other methods (based on the investigation of shadow zones, vertical travel-times, and channel waves).

3. Higher Modes

In Fig. 25a and 25b we show curves of group velocities of the second, third, and fourth modes for the profiles in Fig. 22a. A comparison of Figs. 23 and 25 indicates several advantages of using higher modes:

a) the greater depth of penetration for a given period. To investigate depths up to 300 km with the first mode, we must use periods in the range 200-250 sec, while if we use the third mode periods of 15-20 sec are satisfactory, i.e., observations made by ordinary stations.

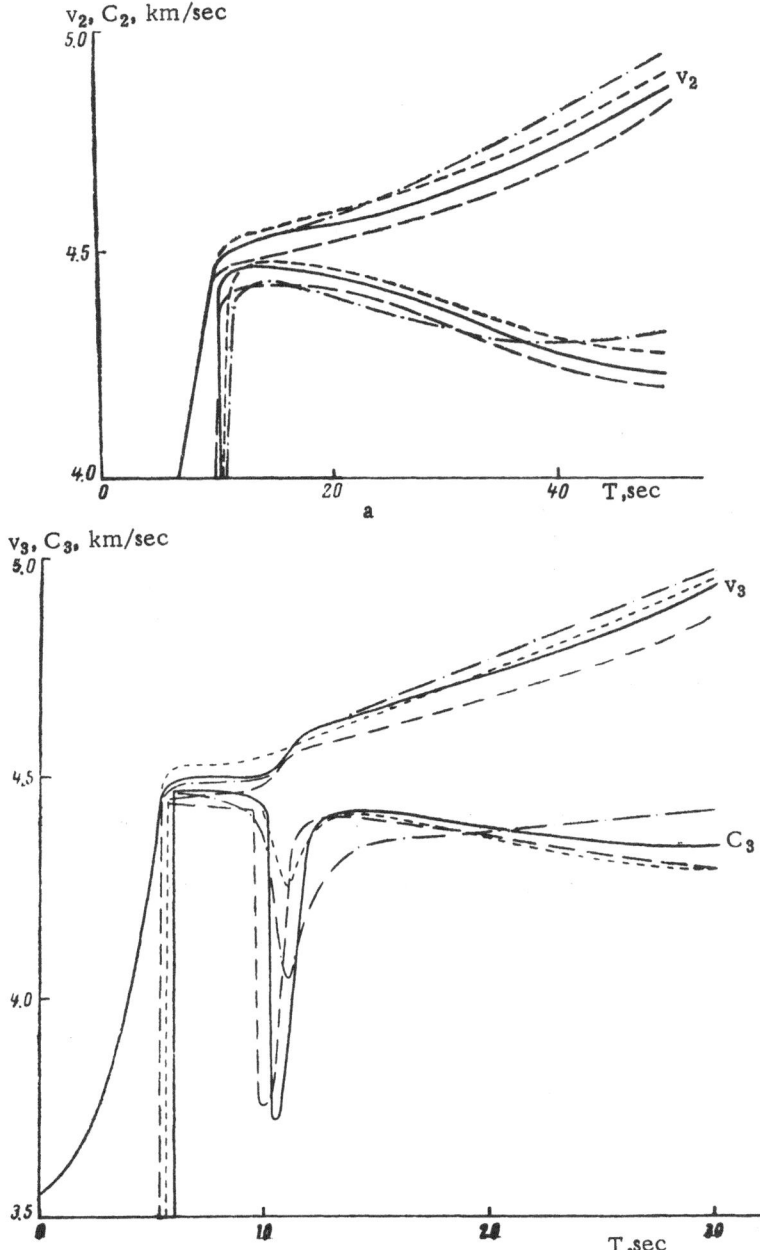

Fig. 25. The dispersion of phase velocities v_k and group velocities C_k for higher modes in models with the velocity profiles of Fig. 22a. a) Second mode; b) third mode.

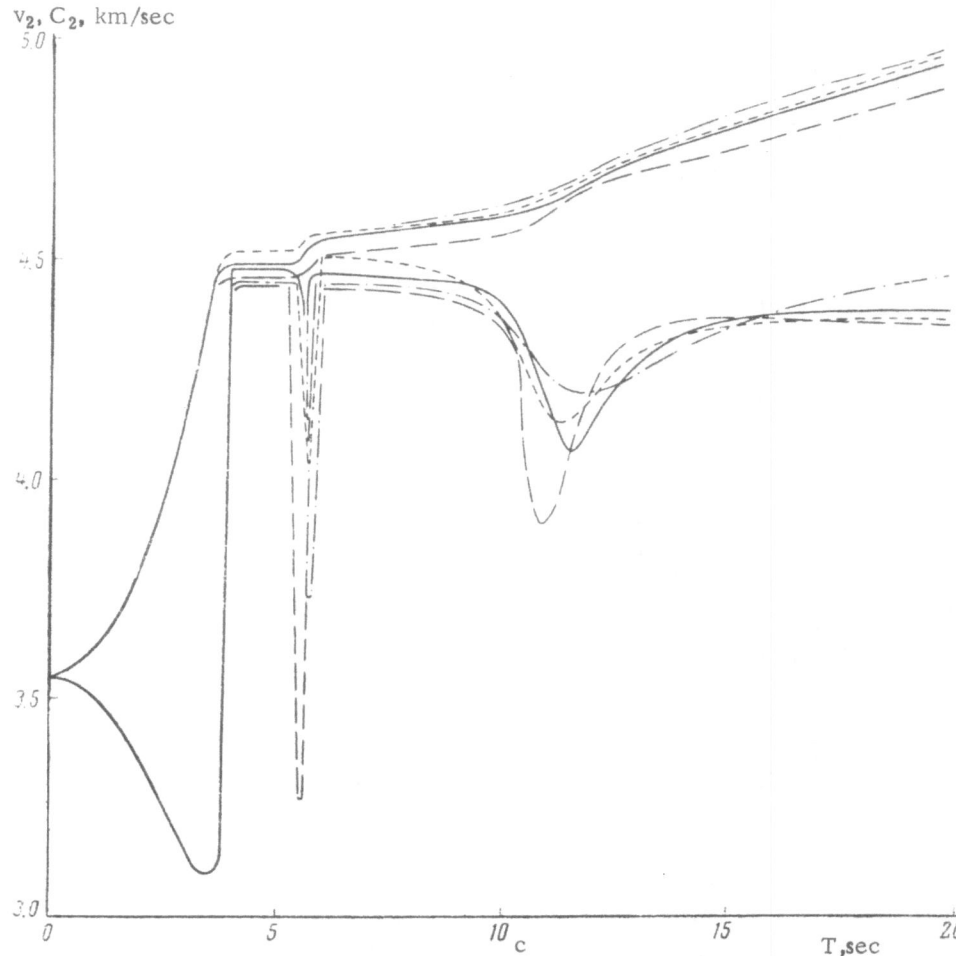

Fig. 25. (continued). c) Fourth mode.

b) a greater resolving power (different profiles show greater differences, and there is a closer relation between the details of a profile and the dispersion curves). For continental models such a relation is particularly clear starting with the third mode (Fig. 25b), for oceanic models the first mode is satisfactory.

We now consider some features of the dispersion of higher modes in various models of the mantle with a continental crust (Fig. 26).

A) A Smooth Increase of Velocity with Depth

For models of this type (indicated by dots and dashes in Fig. 26), a characteristic feature is the smooth shape of the curves for $v_k(T)$ and $C_k(T)$ at periods T sec $\approx \frac{1}{3}$ H km for $k = 2$, T sec $\approx \frac{1}{6}$ H km for $k = 3$; T sec $\approx \frac{1}{7}$ Hkm for $k = 4$, i.e., at the transition from waves in the crust to waves in the mantle.

When T increases $v_k(T)$ increases smoothly.

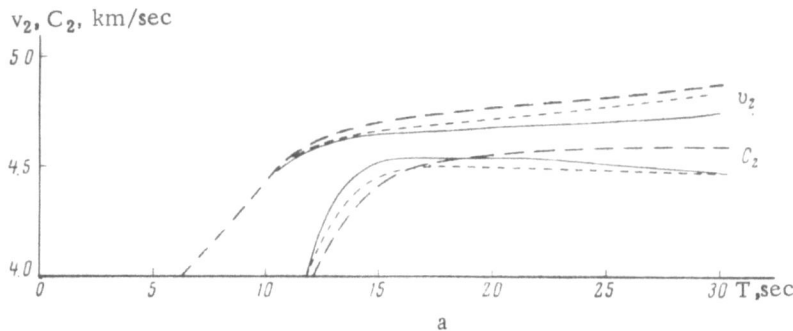

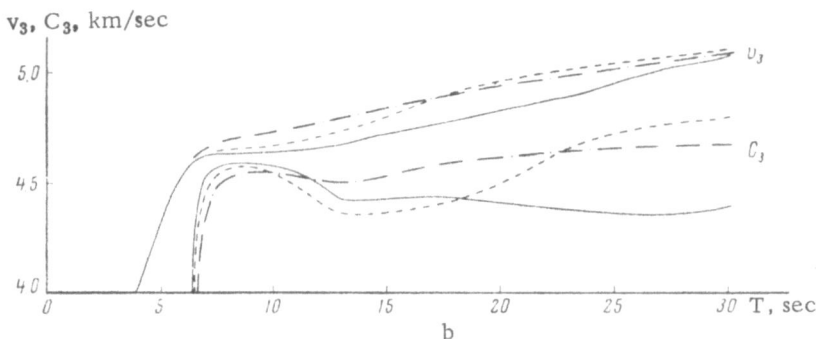

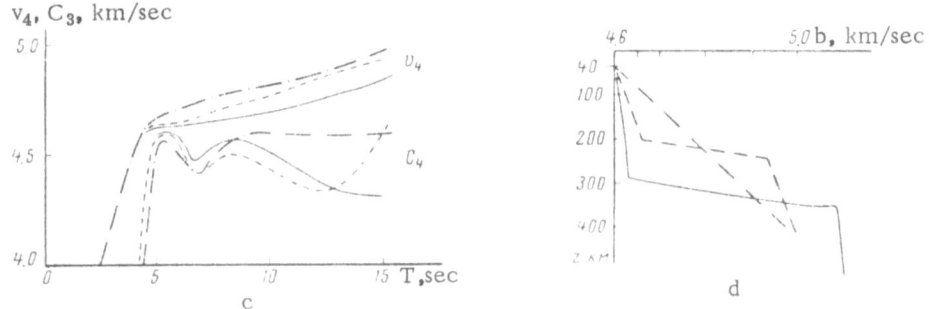

Fig. 26. The dispersion of phase and group velocities of higher modes in the models of the mantle shown on the right at the bottom. a) k = 2; b) k = 3; c) k = 4.

The group-velocity extrema are weak, the velocity drop being of the order of 0.1-0.15 km/sec for $k \leq 4$. The position of the extrema depends on the thickness of the crust and on the mean velocities in the crust, and also on the velocity gradient in the mantle.

B) A Smooth Increase in Velocity on the Background of Which is a Zone of Raised Gradient

Phase velocities in models of this type (Fig. 26, the continuous and dashed curves) are smaller for small T and larger for large T than in models A. The group velocity behaves in the opposite fashion.

For models A and B the phase-velocity differences are not too large (less than 0.15 km/sec for the first three modes); group-velocity differences are somewhat larger — up to 0.2 km/sec for the third mode. These models are distinguished by a relatively large drop in velocity (0.2-0.25 km/sec) between the maximum and minimum for $k \geq 3$. Fig. 26 shows clearly that the effect of the zone of raised gradient on the dispersion increases as the mode number increases.

C) A Mantle with a Waveguide (Fig. 6C, 2)

Here specific features appear [13], the general form of which are described in Section 3, Paragraph 8.

We consider models with continental and oceanic crusts separately.

A Continental Crust. In zones of principal minima (related to the crust), the group-velocity dispersion is approximately the same as for models A and B. For long periods there are small steps on the phase-velocity curves, and sloping sections and minima on the group-velocity curves. These features appear clearly in the third and higher modes. We consider them in more detail for the third mode in the continental-crust model I(H = 30 km) on a Gutenberg 2 mantle. We see that, when they leave the principal-minimum zone (T = 4 sec), the curves for $v_3(T)$ and $C_3(T)$, in contrast to the curves for models A and B, are very steep (with (dv_3/dT up to 0.3 km/sec^2 and dC_3/dt up to 4.6 km/sec^2). Moreover the phase velocity increases very slowly for periods between 4 and 8 sec (on the average $dv_3/dT < 0.01$ km/sec^2), and C_3 is almost constant and differs from $b(\overline{z})$ by not more than 0.03 km/sec.

At the end of this interval $v_3(T)$ approaches very close to the sloping section of $v_2(T)$, almost reaching it for 0.25 sec, and then increases sharply to 0.1-0.12 km/sec (dv_3/dT reaches a value of 0.1 km/sec^2 in this section). This increase corresponds to a sharp local minimum $C_3(T)$ with a value of 0.7-0.8 km/sec for periods of 8-9 sec. The curve for $C_3(T)$ in the local-minimum section intersects the section of sharp increase of the curve for $C_2(T)$ where it leaves the principal-minimum zone, so that the second and third modes correspond to the same group velocities in this short interval.

For periods longer than 10 sec, the third mode depends on deeper layers and only depends slightly on the presence of a waveguide.

The behavior of higher modes is on the whole similar, with the difference that they have more steps on the $v_k(T)$ curve and correspondingly more sloping sections and minima on the curves for $C_k(T)$. With increasing k the velocity drop to a minimum for a fixed mode is larger (the width of the minimum decreasing) and it is displaced in the direction of shorter periods.

Calculations show that when the waveguide becomes deeper all the above features of $C_k(T)$ curves become less marked and are appreciable only for longer periods.

When the crust becomes thinner all the minima are displaced towards shorter periods; at the same time their width decreases and their depth increases, and the local minimum of $C_k(T)$ becomes sharper and deeper.

Oceanic Crusts. In models with thin (oceanic) crusts supplementary calculations show that the same features described above are noticeable for the second mode. The sloping section of the curve for $v_2(T)$ occurs for periods in the range 1.5-7 sec, and the step in the curve $v_2(T)$ and the corresponding minimum of $C_2(T)$ occur for periods of 6.5-9 sec. The minimum is less clearly marked than for $C_3(T)$ in a continental crust (its value about 0.3 km/sec). The minima of higher modes are sharper and narrower than for $C_3(T)$

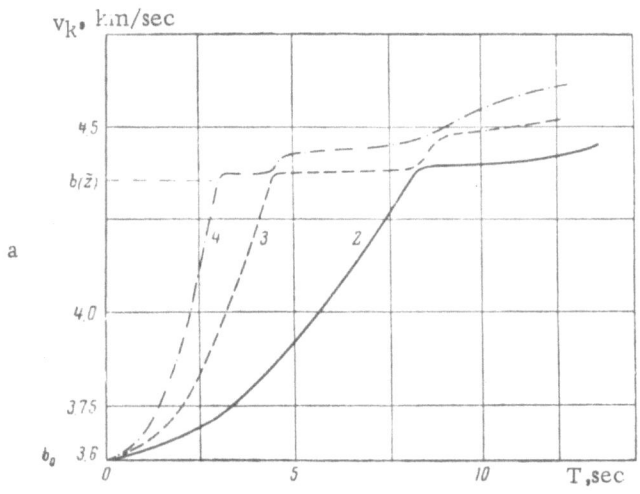

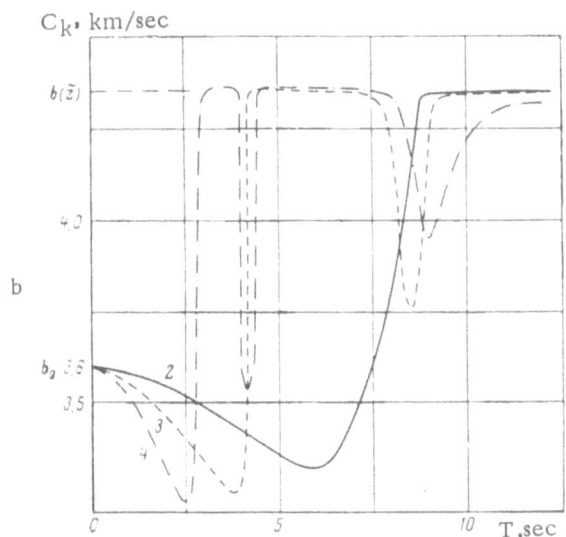

Fig. 27. Dispersion of higher modes in a mantle with a waveguide. a) Phase velocities; b) group velocities; the numbers
on the curves are mode numbers.

but they are located at periods shorter than 2 sec. Hence in models with an oceanic crust the features of the
dispersion under consideration are considerably weaker than in models with a continental crust.

On the whole, judging by the data we have described, the use of higher modes for the investigation of
the structure of the mantle appears to be very promising. There is a clear correlation between definite elements of the profile (the transition to the mantle, waveguides, zones of low and high velocity gradients) and
the dispersion curves of $v_K(T)$ and particularly $C_K(T)$. In this sense waveguides are the most characteristic;
they correspond to steps in the phase-velocity curves and to alternating sloping sections and minima of the
group-velocity curve.

In the realization of the above possibilities, difficulties unavoidably arise in the detection of the various
modes in records, especially in those time intervals in which their periods are close to one another.

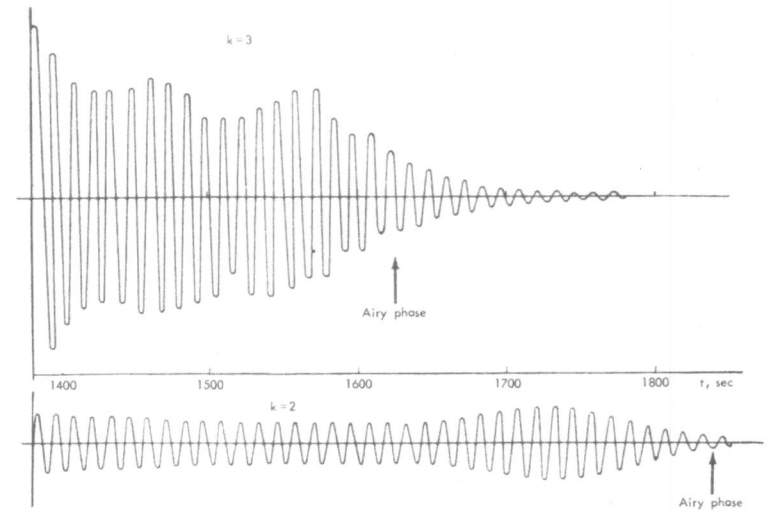

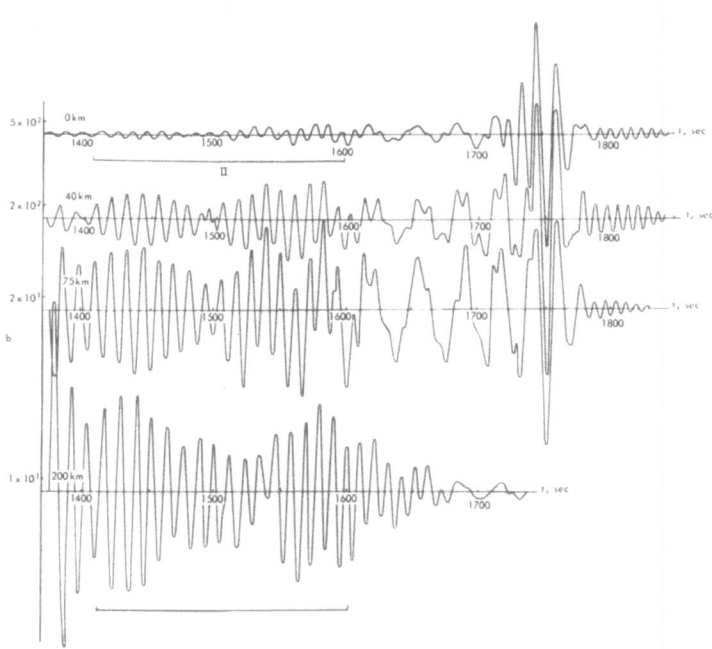

Fig. 28. Theoretical seismograms of Love waves at a distance of 6000 km from the source; the absorption is approximately taken into account; the time is measured from the instant the source starts to act; vertical arrows indicate the beginning of the second and third Airy phases. a) Seismograms of the second and third modes in model II (with waveguide and continental crust), focus depth 75 km; b) complete seismograms for model II, foci depths 0, 40, 75, and 200 km;

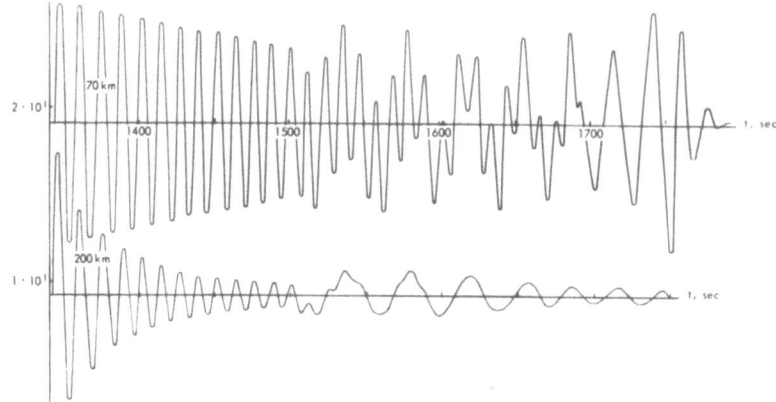

Fig. 28 (cont). c) complete seismograms for model I (without wave-guide), foci depths 75 and 200 km.

§9. The Theoretical Wave Picture

The dispersion curves $C_k(T)$ and the frequency characteristics $\overline{V}_k(T, h)$ can be used to estimate possible conditions for the recording of Love waves penetrating the mantle.

The wave that can be best separated from the other waves and from other modes of Love waves is the long-period part of the first mode.

By comparing Figs. 22 and 23, we easily find that the first mode penetrates the mantle for periods longer than 100 sec (for which the velocity v_1 becomes larger than at the top of the mantle). For $100 < T < 250$ sec, the group velocity is almost stationary ($\sim$4.25-4.5 km/sec), so that long-period quasipulse vibrations are generated. It is propagated in the depth interval between the free surface and the region of raised velocity gradient, and from Fig. 2 we see that the penetration increases continuously as T increases.

The theoretical velocities, the spectrum, and the quasipulse nature of the fundamental mode for $T > 100$ sec agree with the properties of a wave detected on records obtained by long-period apparatus and named a G wave in honor of B. Gutenberg [33, 35, 41, 48, 50]. Observations of strong earthquakes with special apparatus show that this wave is so weakly damped that it can be reflected several times through the earth. This low damping could be due to the weak absorption of very long periods. Moreover it might be assumed that the damping of Love waves of very short periods is also related to horizontal variations of velocity profiles in the crust and the upper mantle; G waves are mainly propagated in the part of the mantle where strong horizontal variation is very improbable.

Essentially shorter-period vibrations related to higher modes in the mantle precede and are partly super-imposed on the fundamental mode. These modes have a quasipulse nature for models without waveguides; the most intense phases are related to the corresponding group-velocity minima of higher modes (see Fig. 18). These vibrations are propagated in the same depth interval as the fundamental mode; their group velocities reach 4.4-4.5 km/sec and correspond to velocities of Sn waves, i.e., to waves identified as transverse head waves generated at the Mohorovicic discontinuity. The group velocities of various higher modes in this range are very close to one another and so they arrive simultaneously; the separation of modes becomes very difficult; vibrations with frequency spectra and times of arrival differing only slightly but with different phase velocities must be separated.

Vibrations associated with Sn waves are often of complex form and of long duration [49], so that it can be assumed that Love waves play an essential role in their formation.

If there is a sharp transitional zone with high gradient in the interior of the mantle (a model of type B), we can expect more clearly dispersed waves — with group velocities 4.2-4.4 km/sec and periods in the range 8-10 sec ($k = 4$) and 10-15 sec ($k = 3$); these waves are related to local group-velocity minima.

When there is a waveguide in the mantle, the preceding fundamental mode and the available periods shorter than 15-20 sec are of low intensity. This is related to the concentration of energy in the waveguide. The corresponding sloping parts of the curves for $v_k(T)$ and the stationary sections of the curves for $C_k(T)$ describe the so-called "channel waves" [25, 34, 39, 42, 49]. It is essential that these waves be, so to speak, trapped by the layer and that they hardly reach the observation surfaces. On the other hand, the presence of the waveguide gives rise to specific vibrations with quasistationary period and long duration, related to the deep local minima of the group velocity and the corresponding peaks of the curves for $\bar{V}_k(T, h)$ (Fig. 21).

For continental models, the most important of these minima are the long-period minima for the third harmonic — $C_3(T)$ (10-12 sec for $H = 35$ km). Vibrations related to shorter-period local minima of $C_k(T)$ for the fourth and higher harmonics carry very little energy since the corresponding peaks of $\bar{V}_k(T, h)$ are very narrow; because of this they can be rapidly damped by selective absorption and by regional variations in the structure of the core and mantle.

The duration of vibrations in the wave with a period between 10 and 12 sec is determined by the amplitude and width of the minimum $C_3(T)$ (see Section 3, Paragraph 3); it depends on the drop of velocity in the waveguide and on the depth of the waveguide. For model II with the continental crust this section begins at time $t = r$ km/4.5 sec and last until $t = r$ km/3.75 sec, and has a complex interference pattern. For larger t, corresponding to the Airy phase, it has the form of an exponentially damped sine wave. The duration of the Airy phase depends on r; for $r = 6000$ km it lasts about 140 sec, for $r = 1000$ km it lasts about 80 sec.

The maximum intensity of a wave is attained for the normal focus depth. When the focus moves deeper below the crust the intensity first slowly decreases and then, when the focus is below the waveguide axis, it begins to decrease rapidly.

The actual detection of this wave on a real seismogram would be persuasive confirmation of the existence of a waveguide in the upper mantle.

A fixed factor that makes the detection of this wave difficult is its interference with the wave corresponding to the principal minimum of $C_3(T)$ (due to the crust). The periods, amplitudes, and times of recording of these waves are close to one another, but the shapes of the envelopes of the corresponding curves are different (Fig. 28a). This is illustrated in Fig. 28b and 28c, in which the theoretical seismograms are shown of the first three modes for models with waveguides (b) and without waveguides (c). For depths smaller than 70 km, the differences between the over-all seismograms are not great. For a depth of 75 km, and particularly for a depth of 200 km, the presence of a waveguide considerably changes the shape of the envelope.

CHAPTER IV

LOVE WAVES AND THE PARAMETERS OF THE SOURCE

In this chapter we investigate the possibility of determining, by observations on Love waves, the following source parameters: the depth of focus (principally within the boundaries of the crust) — according to the spectra of these waves; the magnitude — according to the amplitudes at various epicentral distances; and the mechanism — according to the azimuthal intensity distribution.

§10. Estimates of the Depth of the Focus

1. Normal Earthquakes

Theoretical Effects. It is shown in [10] that with increasing depth of the focus within the limits of the crust (modelled by one or two homogeneous layers) the spectrum of the fundamental mode of Love waves is considerably weakened for the rather short periods for which this wave is related only to the free boundary and is not an interference wave.

Since Love waves are always interference waves, it is to be expected that this effect would not occur in a homogeneous crust. However the earth's crust is not homogeneous: There is usually a sedimentary or transitional layer in its uppper section (Fig. 6B), and below this the velocity $b(z)$ increases with increasing depth. Hence according to Section 3, Paragraph 6, the intensity of Love waves will weaken with increasing depth according to an approximate exponential law, starting with some critical depth $\overline{z}_k$, such that $v_k(T) < b(z)$ for $z > \overline{z}_k$ (pp. 20-21, Fig. 2).

Because of inhomogeneity of the crust, the critical depth $\overline{z}_k$ for some sufficiently short periods can be much smaller than the thickness H of the crust. This leads to conditions for the appearance of an effect similar to that considered in [10] — the damping of short-period Love waves with increasing depth of the focus in the interior of the crust. This effect is clearly evident in the frequency characteristics of the fundamental mode shown in Section 4 (Fig. 7), and also in the theoretical seismogram for model IIb for various focal depths (Fig. 31).

A similar effect occurs for any other source [45]; for sources other than a simple force the only difference is in the behavior of the intensity of $\overline{V}_k(T, h)$ above the critical depth. For example for a dipole with a moment, which is a better model of a real focus, the variation of the intensity for $h < \overline{z}_k$ also depends on the parameter E/G, which is determined by the spatial orientation of the dipole [see (3.3)]. For very small values of E/G (less than 0.1) the intensity of the fundamental mode initially increases with increasing h and reaches its maximum for $h = \overline{z}_1(T)$.

Calculations. For any fixed dipole orientation and in general for any source, the above effect is only sufficiently clearly marked in the part of the spectrum for which $\overline{z}_k(T)$ is small compared to H — let us say $\overline{z}_1 < 0.1$ H. The graphs of $\overline{V}_{d1}(T, h)$ in Fig. 29a and 29b can serve as an example; these graphs are for model IIB for various values of E/G.

Graphs of this type cannot be obtained experimentally, since foci at different depths can have different mechanisms and different initial spectra. We thus consider how the above effect influences the frequency characteristics of a source-medium system for a single fixed source. It is clear from Fig. 10 that these fre-

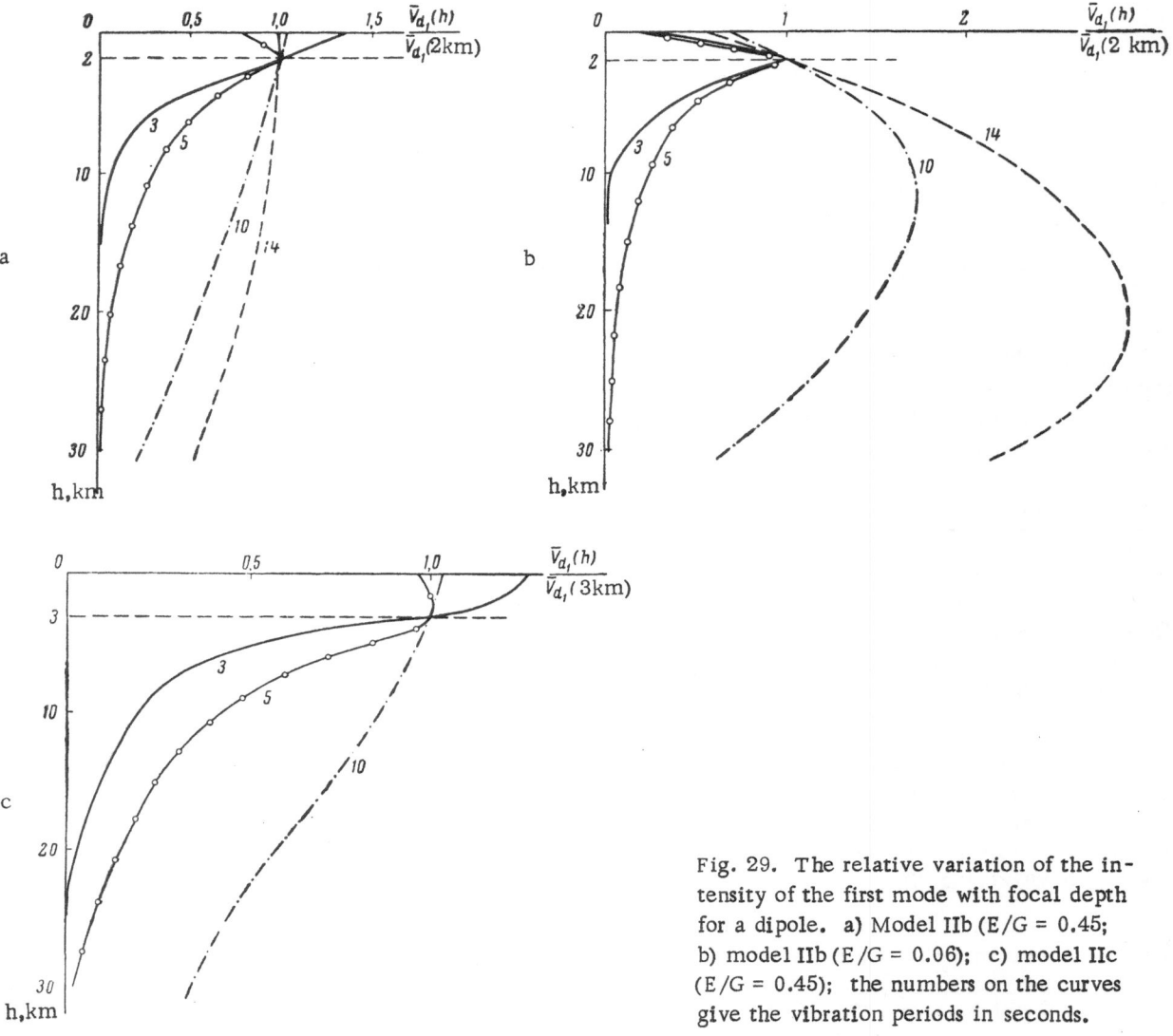

Fig. 29. The relative variation of the intensity of the first mode with focal depth for a dipole. a) Model IIb (E/G = 0.45; b) model IIb (E/G = 0.06); c) model IIc (E/G = 0.45); the numbers on the curves give the vibration periods in seconds.

quency characteristics can in general depend strongly on the source mechanism if the source is not too deep. Independently of the source, however, the frequency characteristic begins to be strongly damped with increasing T as soon as T becomes smaller than the value for which the phase velocity is equal to b(h) [i.e., T decreases so much that $h > z_1(T)$].

As the depth of the focus increases, the dependence of the frequency characteristic on the orientation of the dipole becomes weaker (Fig. 10), and at depths of the order of H/2 the relative distribution of amplitudes in the spectrum becomes practically independent of the source mechanism. Here the basic rule (the relative weakening of short-period sections of the spectrum as the depth of the source increases) is clearly marked in all cases.

For example, in the same crust model IIb the quantity $\overline{V}_{d_1}(T, h)$ satisfies the following inequalities for any dipole orientation:

$$\frac{\overline{V}_{d_1}(5, H/2)}{\overline{V}_{d_1}(10, H/2)} < \frac{1}{3} \frac{\overline{V}_{d_1}(5, h_0)}{\overline{V}_{d_1}(10, h_0)} ,$$

$$\frac{\overline{V}_{d_1}(3, H/2)}{\overline{V}_{d_1}(10, H/2)} < \frac{1}{50}\frac{\overline{V}_{d_1}(3, h_0)}{\overline{V}_{d_1}(10, h_0)},$$

where h_0 is the thickness of the sedimentary layer (2 km). In other words, if the focus is in the middle of the crust, $\overline{V}_{d_1}(T, h)$ decreases at least twice as fast and sometimes fifteen times as fast as when the focus is at the base of the sedimentary layer in the period range from 10 to 5 sec or from 10 to 3 sec.

The Interpretation of Observations. We saw that the short-period components of the spectrum of the fundamental mode of Love waves are strongly damped when the focus is below the middle of the crust.

The period at which this damping begins is mainly determined by the thickness of the sedimentary layer: In the presence of a thick sequence of sedimentary rock of geosyncline type (6-10 km), the damping can begin with periods of from 12 to 15 sec and even longer.

In those relatively rare cases in which there is practically no sedimentary layer, the effect described above is considerably weakened and it depends essentially on the increase of the velocity with depth in the crystalline crust. The presence in the upper part of the crust of a thin layer with a high positive velocity gradient compensates to a considerable degree for the absence of a sedimentary layer. (Fig. 29c).

Since there are clearly periods shorter than 2-8 sec in the spectra of Love waves at epicentric distances that are not too great ($\Delta < 2000$ km), the possibility arises in principle of estimating the focus depth h within the limits of the crust from the damping of the short-period section of the spectrum of the recorded Love waves.

In the investigation of actual earthquake records, the damping of high frequencies in Love waves with increasing h can appear as follows: visually — in the variation of the intensity of short and longer periods; in the steeply decreasing amplitude spectrum in the direction of small periods; in the displacement of the spectrum maximum in the direction of long periods, etc. This is illustrated by examples of real spectra of surface waves for normal earthquakes with a known focus depth [20]. Figure 30 a shows smoothed and maximum-normed spectra of records of surface waves recorded at the Frunze station for several Middle-Asiatic earthquakes ($\Delta = 600$-700 km); Figure 30b shows the same spectra obtained from records taken at the Magadan station of several earthquakes that occurred off the Kamchatka coast ($\Delta = 1100$-1200 km). In both cases the records were obtained with horizontal Kirnos instruments [3]; the spectra were calculated by an electronic computer from numerical records.

It is clear that the spectra for h = 5-10 km differ from those for h = 20-30 km by a distinct displacement of the maximum and of the left (low-frequency) slope in the direction of higher frequencies; for the Far East there is a further difference — the high-frequency slope is not as steep. As a quantitative feature of the spectra we also investigated the area between the curve for $u(p)/u_{max}$ and the horizontal axis in the following frequency intervals: $0 < p < 0.7$ for Middle Asia; $0 < p < 0.4$ and $0.7 < p < 1.5$ for the Far East. For h = 5-10 km the first two areas are smaller and the third larger than for h = 20-30 km. These differences are statistically significant. This preliminary result indicates the possibility of classifying normal earthquakes according to the focal depth by using surface waves. To realize the possibility in practice, we must collect standard spectra for earthquakes with known focal depths or find the statistical distribution of spectra of normal earthquakes.

2. Subcrustal Earthquakes

Because of the sharp increases in velocity at the base of the crust, the depth of penetration $\overline{z}_k(T)$ for a wide range of periods T < 50-60 sec is less than or equal to the thickness of the crust. This is due to the very rapid damping of the short-period part of the spectrum when the vibrations penetrate below the crust. The frequency characteristic $\overline{V}_k(T, h)$ for a fixed depth h > H in the indicated frequency range can be expressed approximately by the formula

$$\exp\left(-\frac{2\pi(h-H)}{Tv_1}\sqrt{1-\frac{v_1^2}{b_M^2}}\right),$$

where b_M is the transverse-wave velocity below the crust. Thus for model II with mantle 1, the ratio $\bar{V}_1(10, h)/\bar{V}_1(20, h)$ (i.e., the relative decrease in the frequency characteristic between periods of 20 and 10 sec) is smaller than its value for h = 30 km by a factor of 30 for h = 75 km, by a factor of 200 for h = 100 km, and by a factor 7000 for h = 150 km. Hence for subcrustal earthquakes there must be a characteristically sharp weakening of periods of from 10 to 15 sec compared to longer periods. This effect must become more marked when the sensitivity of the recording apparatus is higher for longer periods.

3. Features of Higher Modes

An effect similar to that considered in Paragraphs 1 and 2 of this section also occurs for higher modes. It will appear, however, for much shorter periods than in the case of the fundamental mode, since the depth of penetration $\bar{z}_k(T)$ for fixed T increases with the number k (see Section 3). Hence when the modes are superimposed, the higher modes can maximize this effect for the lower modes (including the second mode). This creates known practical difficulties in estimating the focus depth from surface-wave spectra. However when there are surface layers with low velocities in the medium, the fundamental mode in the damped parts of the spectrum (from 10-12 to 2-3 sec) arrives much later in time than the higher modes, so that it can easily be analyzed independently. The long-period part of the fundamental mode (T > 15 sec), needed for comparison, is propagated with high velocities and interferes with the higher modes; they are easily separated, however, since the periods of the higher modes are much shorter (see Section 6).

§11. The Determination of the Magnitude M with the Depth of the Focus Taken into Account

At the present time a universal standard energy characteristic of the intensity of earthquakes is the so-called magnitude M — the logarithm of the maximum displacement or of the velocity at a distance of 100 km from the epicenter. Magnitudes are determined from seismograms of body and surface waves; the method of obtaining them from surface-wave records is described in detail in several articles [5, 19]. It is based on formulas of the type

$$M = \lg\frac{A}{T} - \sigma(r) + c. \tag{11.1}$$

Here A/T is the maximum value of the ratio of the amplitude to the corresponding apparent period (for A some type of total or mean of the amplitudes of the different components is used); $\sigma(r)$ is a calibration function — an empirical law giving the damping of A with increasing epicentral distance r.

The averaging of results from stations with different azimuths and at different distances weakens the effect of the focal mechanism on the quantity $\log(A/T)_{max}$, i.e., the nonuniformity of the radiation from the source of seismic energy, and also the effect of inhomogeneities in the path of propagation. But one essential factor — the focal depth when surface waves are used — is never taken into account except for the recommendation that the procedure described be trusted only for the normal focal depths. Proposals for the correction of the magnitude for focal depth which have appeared in separate articles have not been widely applied [19, 26]. At the same time the decrease of intensity of surface waves when the focus becomes further below the crust has certainly been observed experimentally. For example, by a comparison of magnitudes determined from body waves (for a single earthquake) with magnitudes determined from surface waves the importance of the latter when the focus is deep in the crust or particularly when it is in the subcrustal layer becomes clear [26]. The explanation of this phenomenon is given in Section 4, Section 7, and Paragraph 1 of Section 10. We attempt to give a more detailed estimate of the effect of the focal depth on the magnitude.

1. The Depth of Penetration of Waves Used in the Determination of Magnitude

The records of surface waves obtained by horizontal devices are usually the result of interference of various modes of Love and Rayleigh waves. The predominant periods of vibration for r = 1000 − 6000 km are in the range 7-25 sec [21]. Calculations show that, for focal depths less than 100 km, the maximum amplitude in this period range is related to the fundamental modes. The maximum phase velocities $v_1(T)$ of the first Love wave modes in this period range do not exceed 4.1-4.2 km/sec, i.e., they are much lower than the minimum possible velocity of transverse waves in the mantle. The maximum phase velocities of Rayleigh waves are still lower. Hence for waves of both types the depth of penetration $\overline{z}_1$ is smaller than or equal to the depth H of the Mohorovicic discontinuity (the bottom of the earth's crust). Hence from Section 3, Paragraph 6, for any component of the spectrum of Love (or Rayleigh) waves in the above period range, the intensity $\overline{V}_1(T, h)$ [$\overline{\omega}_1(T, h)$, $\overline{w}_1(T, h)$] decreases as the distance of the focus below the crust increases approximately according to an exponential law:

$$\overline{V}_1(T, h) = \overline{V}_1(T, H)\exp\left[-\frac{2\pi}{Tv_1(T)}\sqrt{1 - \left[\frac{v_1^2(T)}{b^2}\right]}(h - H)\right] \qquad (11.2)$$

where $\overline{V}_1(T, H)$ is the intensity obtained when the focus is at the bottom of the crust and b is the mean velocity in the mantle in the interval between h and H.

Calculations with formula (11.2) show that the quantity $\log\{\overline{V}_1(T, H)/\overline{V}_1(T, h)\}$ is approximately equal to 0.0085 (h - H) for T = 30 sec, and about 0.08 (h - H) for T = 7 sec, where the depths h and H are given in km (Fig. 32).

These estimates give respectively the lower and upper limits of possible variations in the magnitude when the distance of the focus below the bottom of the crust increases. To obtain more accurate corrections we need theoretical seismograms for various focal depths and epicentral distances.

2. Theoretical Seismograms

N. P. Grudeva calculated theoretical seismograms of Love wave modes for model IIb (a low gradient crust with a sedimentary layer on model 2 of the mantle) for r = 1000 km and 6000 km. The depths h of the focus used were 0, 10, and 30 km (in the crust) and 50, 75, 100, and 125 km (below the crust). The source was assumed to be a concentrated force with a uniform spectrum in all frequency intervals. Formulas (1.61-1.62) with $T \approx 2 - 40$ sec were used in the calculations (Airy phases are absent for these periods). Absorption in the medium was roughly taken into account by the introduction of the extra factor $\exp(-\alpha_1 r)$ in which, from the results in [41], $\alpha_1 = 0.017\ T^{-1.42}$ km^{-1}. For periods longer than 10 sec a correction was also introduced for the frequency characteristics of a standard recording device of the SGK type [3]. A part of the resulting seismograms is shown in Fig. 28 and 31.

The maximum value of $\log(A/T)_{z=h} + c_0$, is obtained from the seismograms, where c_0 determines the scale of the seismograms for a given r and is independent of the focus depth.

The difference between these values of ΔM ($\log(A/T)_{z=H} - \log(A/T)_{z=h}$) depends mainly on differences between focal depths and only varies slightly for changes in the epicentral distance. We investigate $\Delta M(h - H)$, where H is the crust thickness. In the crust (h < H) the variation of ΔM is insignificant. For sub-

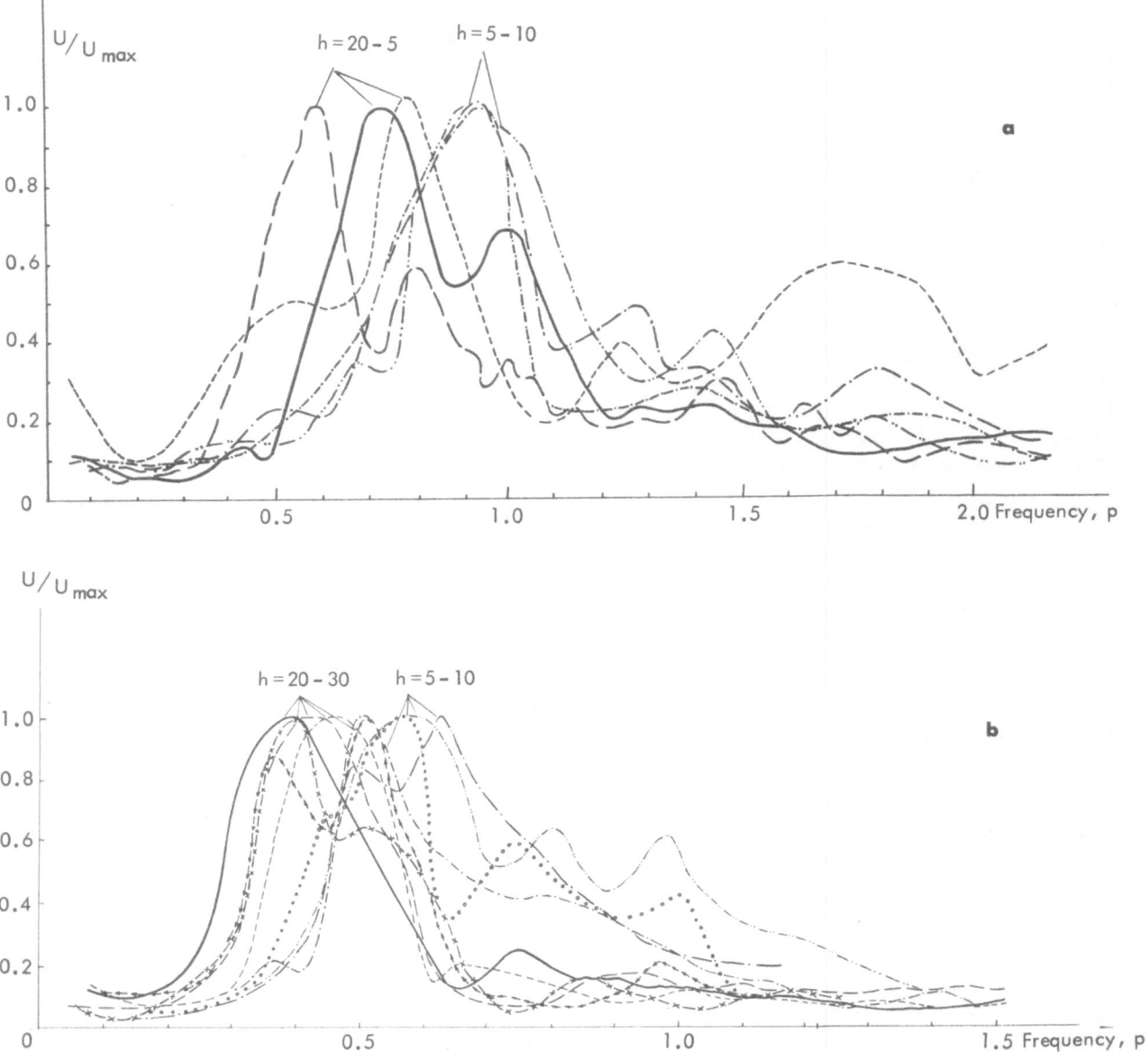

Fig. 30. Smoothed spectra of surface waves in earthquakes. $M = 5\frac{1}{4} - 5\frac{1}{2}$. a) Frunze station, r = 600-700 km; b) Magadan station, r = 1000-1200 km.

crustal earthquakes with r = 1000-6000 km values of $\Delta M(h - H)$ are shown in Fig. 32. A band of values of ΔM is given, determined from seismograms of the fundamental mode (vertically shaded) and from the overall seismogram of the first three modes (shaded with oblique lines). In Fig. 33 the periods corresponding to max log $(A/T)\rho$ are shown in the same notation. There is a strip of values because of scattering due to the variation of epicentric distances and focus mechanisms. The straight line in Fig. 32 shows the limiting values of ΔM for the fundamental mode, calculated in Paragraph 1 of this section.

The divergence of the strips in Fig. 32 and 33 is caused by the fact that, for h - H> 50 km, the maximum amplitudes already belong to the short-period higher modes which are only weakly damped with increasing depth.

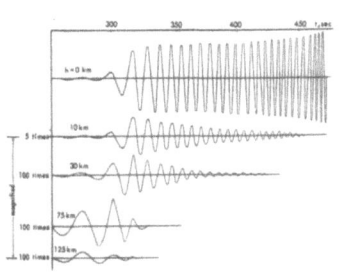

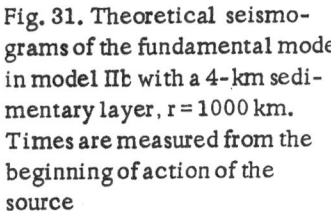

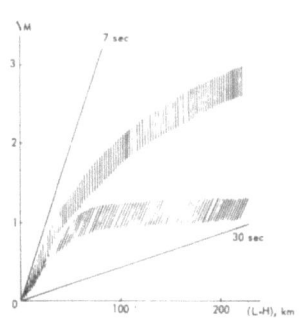

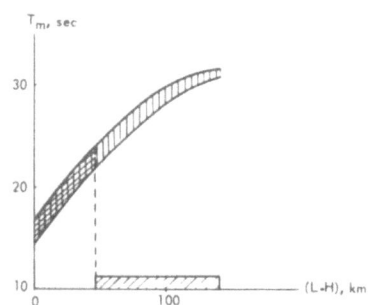

Fig. 31. Theoretical seismograms of the fundamental mode in model IIb with a 4-km sedimentary layer, r = 1000 km. Times are measured from the beginning of action of the source

Fig. 32. Correction to the magnitude ΔM for the depth (h - M) of the focus under the crust, obtained from the fundamental mode or the sum of modes.

Fig. 33. Period of maximum phase as a function of the depth of the focus under the crust.

For the over-all seismogram, ΔM is the correction to the magnitude for the focus depth if M is strictly determined according to formula (11.1) on p. 67. We see that this value of ΔM for foci with h = 100-150 km reaches the value 1.

If the magnitude is determined from vibration with periods in the range 15-30 sec, then the correction is equal to ΔM for the fundamental mode. Although its rate of increase with increasing h is slowed became of the increase of the predominant periods, it reaches 1.5-2 for h ≈ 100-150 km. For long-period devices of the Press-Young type, the strips in Fig. 32 should be displaced towards one another.

A similar effect must exist for Rayleigh waves. In this connection, as has already been noted, the presence of a weakened layer in the mantle with minimum velocity $b(\tilde{z}) \approx 4.35$-$4.40$ km/sec has no important effect on the drop in intensity of surface waves with increasing depth in a continental crust. At sufficiently great depths (h > 100 km) the intensity of the fundamental harmonics of surface waves in the period range under consideration is so weakened that the maximum values of log (A/T) can be attained for the higher modes (especially when r is not too large). This will make itself felt in a decrease of the corresponding periods and an increase in the group velocity in the section of the record with maximum amplitude. Higher modes, which have phase velocities higher than the minimum velocities $b(\tilde{z})$ in the mantle for periods of 10-20 sec, are characterized by deep penetration and much larger H. Hence for higher modes the variation of log (A/T) with increasing h becomes slower. This effect has not yet been investigated quantitatively.

Thus the sharp decrease in magnitudes obtained from surface waves at some focal depth below the crust can be explained in the framework of the theory in Sections 1-3 by simply increasing the velocity at the crust-mantle boundary, without assumptions concerning the existence of an asthenosphere [26].

With certain known reservations, the graph in Fig. 32 can be used to obtain an estimate of corrections to the magnitude for focal depths below the crust. This method can be used when we are sure that the maximum amplitudes on the record correspond to the fundamental mode.

§12. The Azimuthal Intensity Distribution of Love Waves

It follows from (3.1) that the amplitude spectrum of the k-th mode for a fixed frequency p is a function of the azimuth φ with epicenter at the station.

This function depends essentially on the type of source. For a concentrated source, from (3.1) and (3.2) the variation of $U_\varphi^k(\varphi)$ for large r is described by a factor $\sin(\delta - \varphi)$, where δ is the azimuth of the horizontal

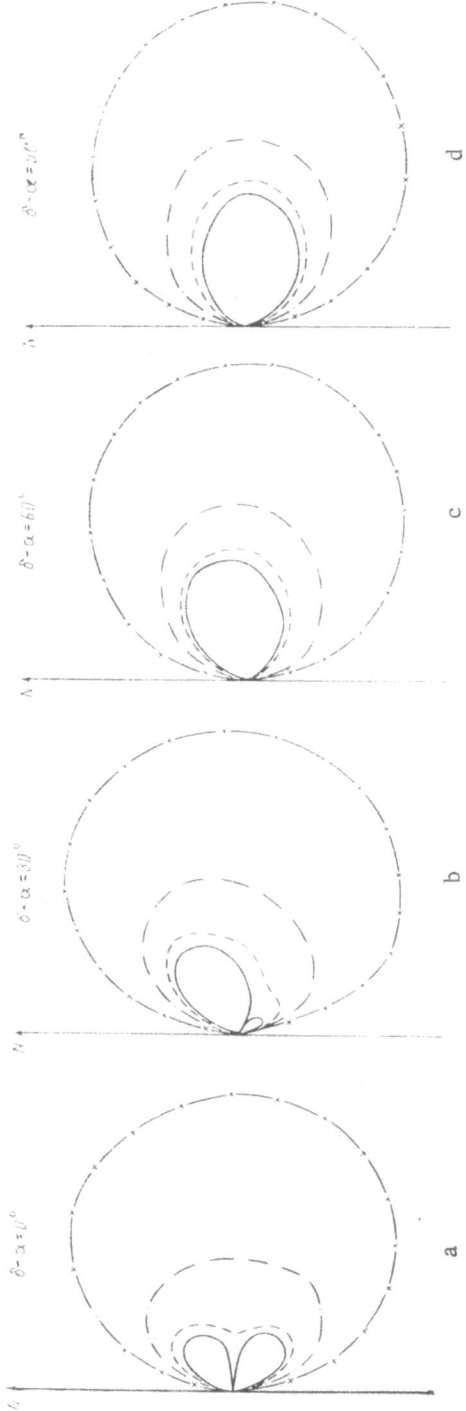

Fig. 34. Vector diagrams of k-th modes of Love waves for various values of the angle $(\delta - \alpha)$ (for similicity $\delta = 0$). The parts of the diagrams to the left of the vertical axis can be obtained by rotating the part to the right about the pole through 180°. The curves with successively increasing radii correspond to $(S/R^2) = 0$; 0.2; 1.0; 5.0.

projection of the force; hence the form of $U_{\phi}^{k}(\varphi)$ is independent of the frequency p, the focus depth h, and the structure of the medium. Figure 34 contains vector diagrams for values of $(S/R)^2$ from 0 to 5.

For a dipole with a moment, the functional dependence described by $U_{\varphi}^{k}(\varphi)$ is of a more complex nature; from (3.1) and (3.3), for this source we have

$$U_{\phi}^{k}(\varphi) \sim \sin(\delta - \varphi)\sqrt{R^2 \cos^2(\alpha - \varphi) + S^2}. \tag{12.1}$$

Here

$$R = |\tilde{V}_k(h)\,\tilde{V}_k(0)|\sin\gamma, \tag{12.2}$$

$$S = v_k/p\,|\tilde{V}_k'(h)\,\tilde{V}_k(0)|\cos\gamma.$$

We recall that α is the azimuth and γ is the dip of the plane of discontinuity.* It is clear from (12.1) that the azimuthal diagram differs from $\sin(\delta - \varphi)$ more when the ratio S/R is smaller. It can be seen from (12.2) that this ratio depends on the mode number k, the frequency p, the focal depth h, the structure of the medium [which determines $v_k(p)$, $\tilde{V}_k(p,h)\,\tilde{V}'(p,h)$], and the dipole parameters γ, α, δ.

From (12.1) we see that the basic nodal plane coinciding with the direction of the force $\varphi = \delta$, is unchanged for a dipole. In the general case ($\alpha \neq m\pi/2$) there is no mirror symmetry relative to this plane (but there is symmetry relative to the epicenter). For values of S/R that are not too small there are only two loops corresponding to the zeros of $\sin(\delta - \varphi)$. Only for S/R → 0 do four loops appear; they begin at the zeros of $\cos(\alpha - \varphi)$; this case corresponds to $\gamma \to \pi/2$ (vertical incidence) or to $\tilde{V}_k'(h) \ll \tilde{V}_k(h)$.

This last case for models with positive velocity gradient for k = 1 can only occur for h → 0; for k > 1 it can occur close to the extrema of $\bar{V}_k(h)$.

If $|\delta - \alpha| \approx \pi/2$ or $3\pi/2$, the second nodal plane practically combines with the first.

Since S/R depends on many variables, there is no reason to describe the behavior of $U_{\varphi}^{k}(\varphi)$ for dipoles in general. For each specific dipole we can easily calculate $U_{\varphi}^{k}(\varphi)$ by using the program described in Section 3. For given values of k, p, and h it gives values of $v_k(p)$, $[\tilde{V}_k(p,h)\tilde{V}_k(p,0)]$, and $[\tilde{V}_k'(p,h)\,\tilde{V}_k(p,0)]$. The function $U_{\varphi}^{k}(\varphi)$ can easily be calculated for any φ when α, δ, and γ are given [(3.1)–(3.3)]:

*Instead of any of the angles δ, α, or γ, we can specify the angle θ between the direction of motion and the vertical. These angles are related by the formula $\cos\theta\,\mathrm{ctg}\,\gamma = -\cos(\delta - \alpha)$.

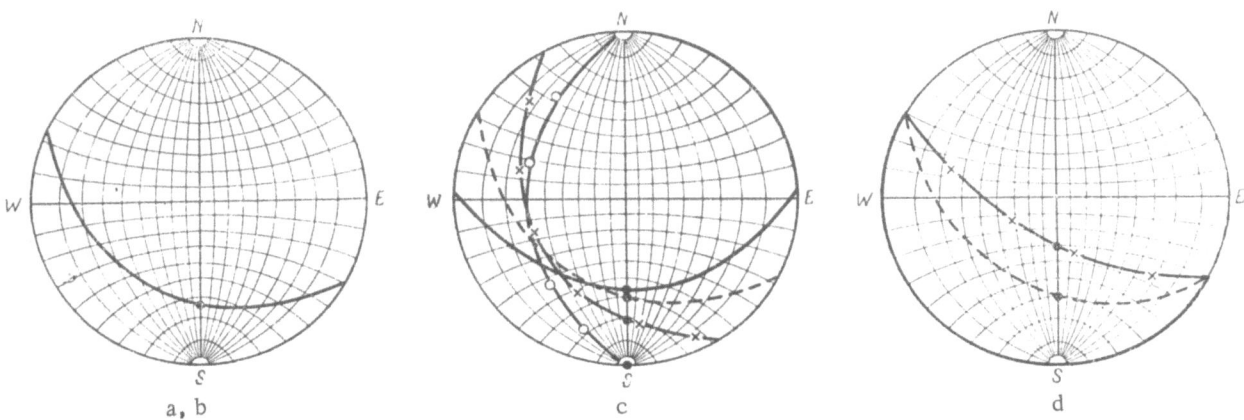

a, b   c   d

Fig. 35. Dipole orientations corresponding to Fig. 36.

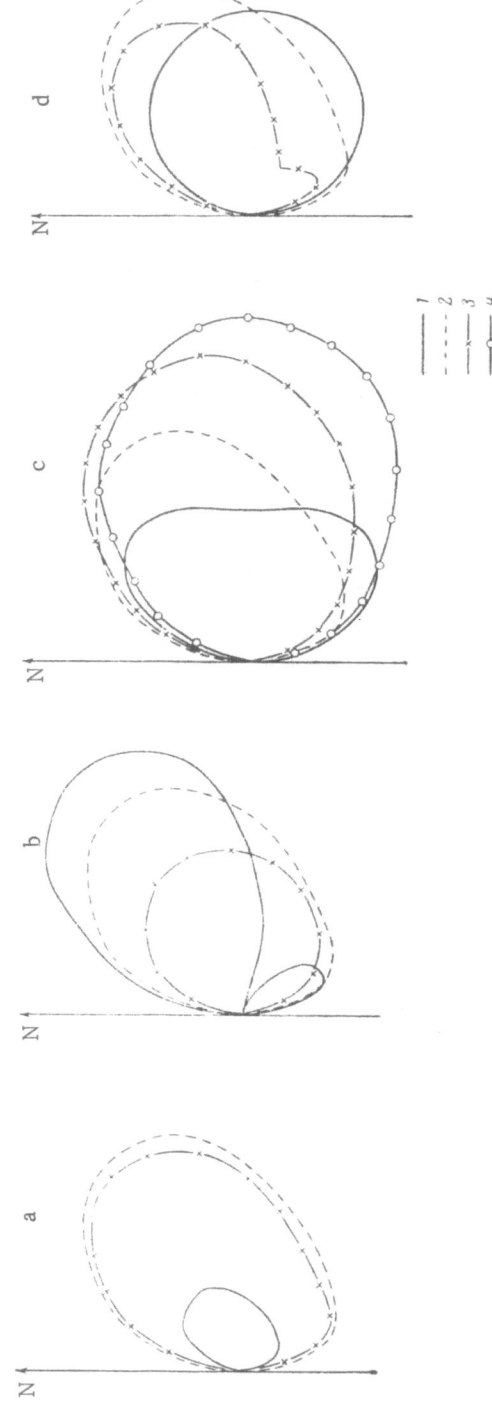

Fig. 36. Vector diagrams of the intensity of the fundamental mode of Love waves (for simplicity $\delta = 0$). The part of the diagram to the left of the vertical axis is obtained by rotating the part on the right through 180° about the pole. a) Fixed parameters: $\gamma = 30°$; $(\delta - \alpha) = 30°$; $h = 15$ km; periods; 1) $T = 5$ sec; 2) $T = 10$ sec; 3) $T = 14$ sec; b) fixed parameters: $\gamma = 30°$, $(\delta - \alpha) = 30°$; $T = 14$ sec; focus depths: 1) $h = 0$; 2) $h = 15$ km; 3) $h = 30$ km; c) fixed parameters: $\gamma = 30°$, $h = 15$ km; $T = 14$ sec; azimuths: 1) $(\delta - \alpha) = 0$; 2) $(\delta - \alpha) = 30°$; 3) $(\delta - \alpha) = 60°$; 4) $(\delta - \alpha) = 90°$; fixed parameters: $(\delta - \alpha) = 30°$; $h = 15$ km; $T = 14$ sec; dip; 1) $\gamma = 0$; 2) $\gamma = 30°$ 3) $\gamma = 60°$.

As an example we consider the azimuth diagrams for the fundamental mode of Love waves in model IIb (a low-gradient crust 30 km thick including 2 km of sedimentary layers).

To separate the effect of the various parameters we vary each of them in turn. We can take $\delta = 0$ (diagrams for other values of δ can be obtained by rotating the graph through the appropriate angle). In Fig. 35 we show the orientations of the dipoles for which calculations were performed: Figure 36 shows the corresponding vector diagrams.

In the first example (Fig. 35a; 36a) the period is varied. Variation of T has almost no effect on the nature of the azimuth diagram (since h is not too small).

In the second example (Fig. 35b; 36b) the focal depth is varied. For h = 0 secondary loops appear with maximum intensity in the regions $\varphi = (\pi + \delta - \alpha)$ and $\varphi = \delta - \alpha$ while supplementary regions of minimum intensity appear close to $\varphi = \pi/2 + (\delta - \alpha)$ and $3\pi/2 + (\delta - \alpha)$. When h increases the supplementary loops disappear.

In the third example (Fig. 35c; 36c) the angle $(\delta - \alpha)$ is varied. The variation of $(\delta - \alpha)$ has some effect on the polar diagrams in the region of their maxima. When $(\delta - \alpha)$ increases from 0 to $\pi/2$, the diagrams become less uniform. When $\delta - \alpha = m\pi/2$ there is mirror symmetry relative to two axes: $(\varphi - \delta) = 0; \pi/2$.

In the fourth example (Fig. 35d; 36d) the dip γ of the plane of discontinuity is varied. For small γ the $U^1_\varphi(\varphi)$ diagram is similar to the diagram for a concentrated force; as γ increases this symmetry disappears and, starting with certain sufficiently large values of γ, secondary loops appear.

The examples in Fig. 36 show clearly that, for dipoles with moments, the appearance of a secondary nodal plane (in addition to that coinciding with the direction of motion) is possible only in relatively rare cases [very small focus depths or a vertical plane of discontinuity for small angles $(\delta - \alpha)$].

This permits the use of observations of Love waves at various azimuths in the determination of the azimuth of motion in the focus, i.e., the angle δ. A preliminary determination of the complete component U_φ by the decomposition of the recorded vibrations into Love and Rayleigh waves is necessary [29].

THE INFLUENCE OF SPHERICITY,
VARIATION OF THE PARAMETERS, AND ABSORPTION

(This appendix contains results obtained after the publication of the Russian edition of this book.)

§13. The Influence of the Sphericity of the Earth

At great distances from the source and for high phase velocities, the approximation of the earth by a plane model is too coarse. Calculations by other methods [51, 52, 53] show that the effect of sphericity is particularly strong for higher modes, in which it appears even at periods of 15-20 sec and shorter.

1. Theory

As a model of an elastic earth with a liquid core we use a perfectly elastic spherical shell, the internal and external surfaces of which (with radii R and R_0) are free from tangential stresses. We use spherical coordinates — the radius l, the lattitude θ, and the longitude φ. Let forces $F(l, \theta, \varphi, t)$ similar to the forces (1.7)–(1.9) act in some region of the shell: The projections of F on the coordinate axes are

$$F_l = F_l(l,\theta,t)$$
$$F_\theta = F_T(l,\theta,t) \cos(\delta - \varphi) \tag{13.1}$$
$$F_\theta = F_T(l,\theta,t) \sin(\delta - \varphi) .$$

Methods similar to those used in Section 1 show that, at large distances θ from the source, the principal part of the disturbance consists of surface waves. The Rayleigh waves are polarized in the $l\theta$ plane and are proportional to the components F_l and F_θ of the force. The Love waves have only an azimuthal component U_φ, and are proportional to the component F_φ of the force at the source.

At sufficiently large frequencies p, we obtain the following expression for the stationary displacement of a point with coordinates l, θ, φ, in a Love wave passing a distance $\theta_1 \cdot l$ * through the sphere and thus traversing the anti-epicenter and epicenter S times (S = 0, 1, 2, . . .):

$$U_\varphi(l, \theta, \varphi, p) = \sum_{k=1}^{k_L(P)} \frac{D_{kL} \widetilde{V}_k(\xi_k, p, l)}{\sqrt{\xi_k |\sin\theta_1| l}} \exp\left[-i\xi_k l\theta_1 + i(\pi/2)S\right] \tag{13.2}$$

$$D_{kL} = l^{-i\pi/4} \sqrt{2\pi^3} \int_{R_0}^{R} \frac{l^2}{R^2} \widetilde{F}_T(\xi_k, p, l) \sin(\delta - \varphi) \widetilde{V}_k(\xi_k, p, l) \, dl . \tag{13.3}$$

These formulas hold for all sufficiently large θ_1 (such that $\theta_1 l\xi_k \gg 1$) such that $\theta_1 \neq n\pi$. Here $\widetilde{F}_T(\xi_k, p, l)$ is as before the space-time spectrum of the source defined as in (1.11), $\widetilde{V}_k$ is the k-th characteristic function and ξ_k^2 the k-th characteristic values of the boundary-value problem

---

*The angles θ_1 and θ are related by the equation $\theta_1 = \theta + S\pi$ for even S; $\theta_1 = (S + 1)\pi - \theta$ for odd S.

$$\frac{d}{dl}\left[u\left(\frac{dV}{dl} - \frac{V}{l}\right),\right] + \frac{3u}{l}\frac{dV}{dl} + \left(p^2\rho - u\frac{\xi^2 R^2}{l^2}\right)V = 0 \qquad (13.4)$$

$$R_0 < l < R$$

for

$$u\frac{dV}{dl} = 0, \qquad \text{with} \qquad l = R_0, R. \qquad (13.5)$$

The functions $V(l)$ and $u(dv/dl)$ are continuous for all l inside the shell. The substitutions $V = ly$ and the multiplication of all terms by l^3/R^4 converts (13.4) into the form

$$\frac{d}{dl}\left[u\frac{l^4}{R^4}\frac{dy}{dl}\right] + \left(p^2\rho - u\xi^2\frac{R^2}{l^2}\right)y\frac{l^4}{R^4} = 0 \qquad (13.6)$$

$$u\frac{dy}{dl} = 0 \qquad \text{with} \qquad l = R_0, R. \qquad (13.7)$$

The functions y and $u(dy/dl)$ are continuous for all l inside the spherical shell.

The problem of finding the characteristic functions $\tilde{y}_k$ and characteristic values ξ_k (13.6-13.7) for given p and k is similar to that considered in Section 2 and is solved by the same method. We therefore only enumerate the fundamental steps of the solution, referring to Section 2 when necessary.

The problem (13.6-13.7) is solved by the exhaustion method [7]. To any solution we associate a corresponding continuous exhaustion function $\theta(l)$:

$$\theta(l) = -\text{arctg}\ \frac{ul^4\dfrac{d\tilde{y}}{dl}}{R^4\tilde{y}}. \qquad (13.8)$$

From (13.6) and (13.8) we obtain the following equation for θ:

$$\frac{d\theta}{dl} = \frac{R^4}{ul^4}\sin^2\theta + \left(p^2\rho - \frac{u\xi^2 R^2}{l^2}\right)\frac{l^4}{R^4}\cos^2\theta \qquad (13.9)$$

on the interval $[R_0, R]$. The boundary conditions for θ are

$$\theta(R) = 0 \qquad (13.9')$$

$$\theta(R_0) = -(k-1)\pi. \qquad (13.9'')$$

The number k in (13.11) coincides with the number of the characteristic value.

The characteristic values ξ_k^2 satisfying (13.9) with a given value of k are obtained by a trial-and-error method. We integrate (13.9) from R_0 to R with the initial condition (13.9'') and calculate $\theta(R)$. We choose a value of ξ^2 for which $\theta(R) = 0$. This value is ξ_k^2.

The method of integration and the selection of ξ^2 is similar to that described in Section 2.

To find the characteristic function $\tilde{y}_k(l)$, we introduce a new independent function $r_k(l)$ by the equation

$$(13.10)$$

$$\frac{ul^4\dfrac{d\tilde{y}_k}{dl}}{R^4} = -r_k\sin\theta_k, \qquad \tilde{y}_k = r_k\cos\theta_k.$$

From (5.6) and (5.10) we obtain the following equation for r_k:

$$\frac{dr_k}{dl} = +\frac{1}{2} r_k \sin 2\theta_k \left[-\frac{R^4}{ul^4} + \left(p^2 \varrho - \frac{u\xi^2 R^2}{l^2} \right) \frac{l^4}{R^4} \right] . \tag{13.11}$$

When the characteristic value ξ_k^2 has been found, we solve (13.9) and (13.11) simultaneously and calculate $r_k(l)$, $\widetilde{y}_k(l)$ and $d\widetilde{y}_k/dl$. A more accurate characteristic value can be obtained by using the following numerical formula from the calculus of variations:

$$\xi_k^2 = \frac{p^2 L_1 - L_3}{L_2} \tag{13.12}$$

where

$$L_1 = \int_{R_0}^{R} \varrho \, \frac{l^4}{R^4} \widetilde{y}_k^2 \, dl; \qquad L_2 = \int_{R_0}^{R} u \, \frac{l^2}{R^2} \widetilde{y}_k^2 \, dl; \qquad L_3 = \int_{R_0}^{R} u \, \frac{l^4}{R^4} \left(\frac{d\widetilde{y}_k}{dl} \right)^2 dl . \tag{13.13}$$

We now calculate the phase velocity $v_k = p/\xi_k$ and the group velocity $C_k = dp/d\xi_k$:

$$\nu_k = \left(\frac{p^2 L_2}{p^2 L_1 - L_3} \right)^{1/2} \tag{13.14}$$

$$C_k = \frac{L_2}{L_1 \nu_k} . \tag{13.15}$$

The spectral amplitude $\overline{V}_k(l, R, p)$ of the k-th mode of the Love wave on the surface $l = R$ generated by a point-pulse force at depth l is calculated from the formula

$$\overline{V}_k(l, R, p) = \frac{l^3 \widetilde{y}_k(l) \widetilde{y}_k(R)}{2R^3 \sqrt{\xi_{kL} L_2}} . \tag{13.15'}$$

The program for the solution of the above problem on a computer is very similar to that described in Sections 2 and 3; it uses the same input and prints out the same information concerning the solution.

2. Calculations

In Fig. 37-39 we show a comparison of numerical results for a spherical and a plane model with identical profiles corresponding to the results obtained by Gutenberg and Birch [57]. Figure 37 shows the phase and group velocities of the fundamental mode. The effect of sphericity is marked for T > 70 sec and leads to a vertical displacement of the curves for $v_1(T)$ and $C_1(T)$ amounting to 0.1 km/sec for T = 100-500 sec. Phase velocities of higher modes (k = 2, 3, 4) are shown in Fig. 38a. Sphericity has an influence here even for periods of 10 sec and less. Thus for the third mode the phase velocities differ by 0.1 km/sec for T = 6-10 sec and by 0.3 km/sec for T = 50 sec. Figure 38b shows group velocities of higher modes. The main features of the curves for the two models are similar. The effect of a waveguide, however, which leads to stationary values and local minima of the velocity as shown in Sections 3 and 8, is much less clear in the spherical model. The effect of sphericity is well explained by ray theory; the introduction of spherical symmetry is equivalent to an increase of the velocity gradient in the plane model [55].

A comparison of frequency characteristics for the two models is shown in Fig. 39; the case of a surface source is illustrated. It is clear that the amplitudes of long-period waves are larger in the spherical model (Fig. 39a). The example of the third mode shows (Fig. 39b) that the resonance of the "crust-waveguide" system is weakened (at periods in 9-13 sec range) and that the energy is not strongly concentrated in the waveguide (at periods of from 5.5 to 9 sec) in the spherical model.

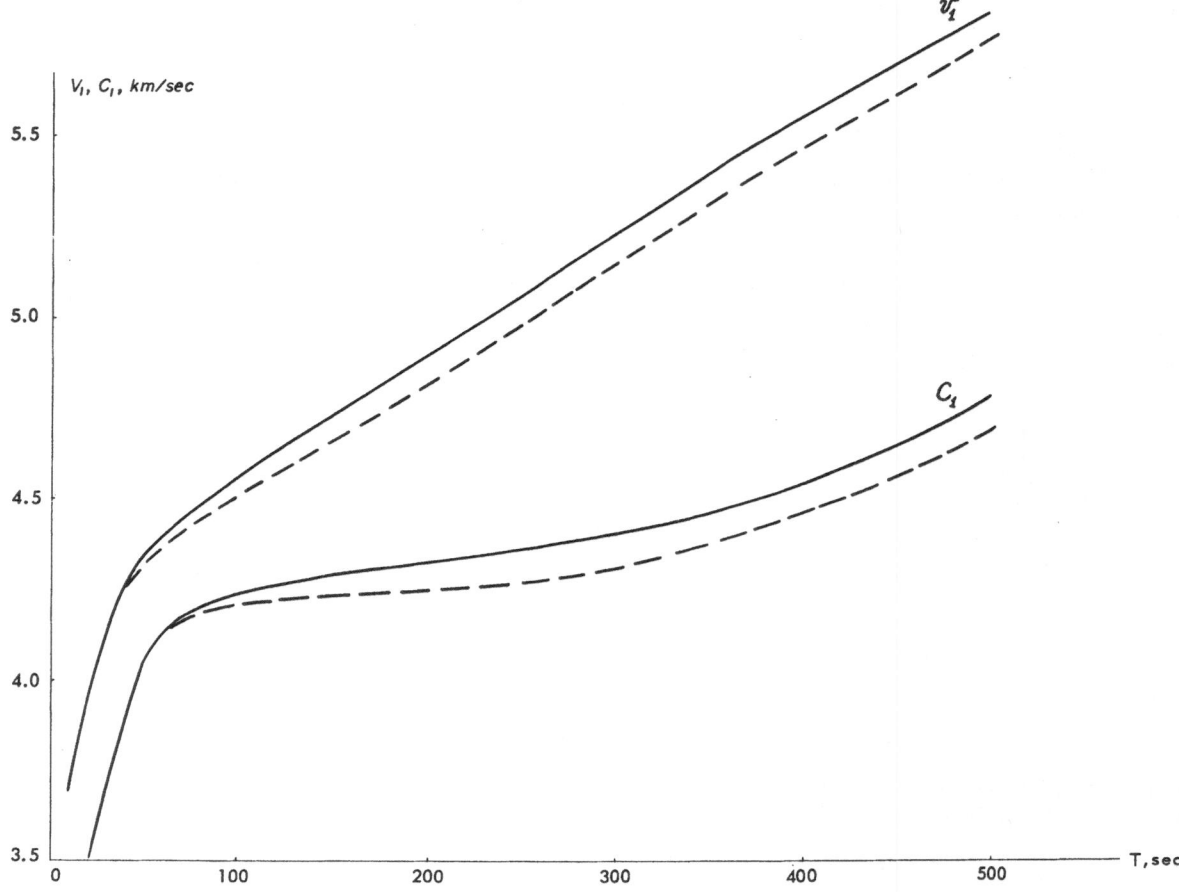

Fig. 37. Phase and group velocities of the fundamental mode. Continuous curve) spherical model; dotted curve) plane model.

§14. Derivatives of the Phase and Group Velocities with Respect to the Parameters for the Section

In the correct formulation of the inverse problem, it is very important to know how to predict the effect of local variations (with depth) of the profile parameters (the transverse-wave velocity b and the density ρ) on the dispersion curves of the phase and group velocities of the various modes. Such a prediction is possible if, for the reference profiles, we know the partial derivatives of the phase and group velocities with respect to the parameters b and ρ in the separate layers as functions of the vibration period [57, 58]. A knowledge of these derivatives is also directly useful in the solution of inverse problems, since it leads to a considerable decrease in the time of calculation.

A method of calculating the disturbances in the dispersion curves due to small variations in the profile parameters was obtained by Jeffreys [56]. This method was used in [57] and [58] in the calculation of surface waves in a system of homogeneous layers. We give below a short description of a method of calculating derivatives that can be used with our apparatus. We confine ourselves to the spherical problem.

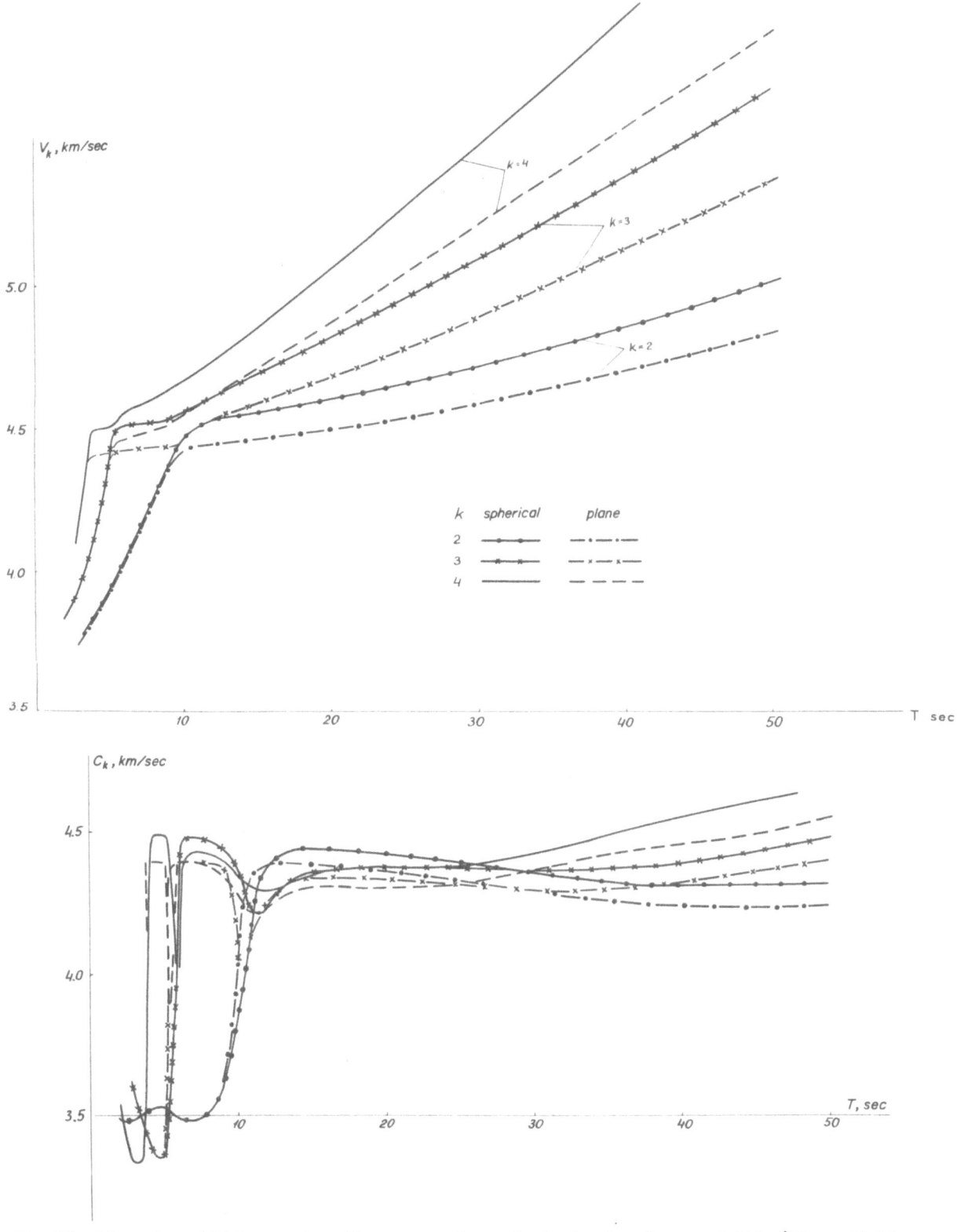

Fig. 38. Dispersion of higher modes of Love waves in a spherical and a plane model II of the earth (the continuous and dotted curves respectively): a) Phase velocities; b) group velocities.

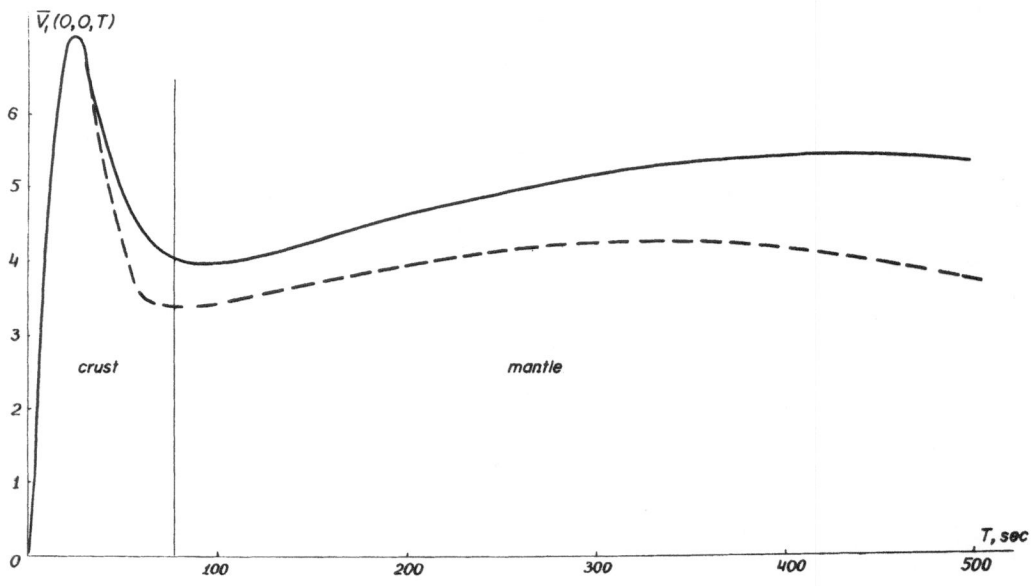

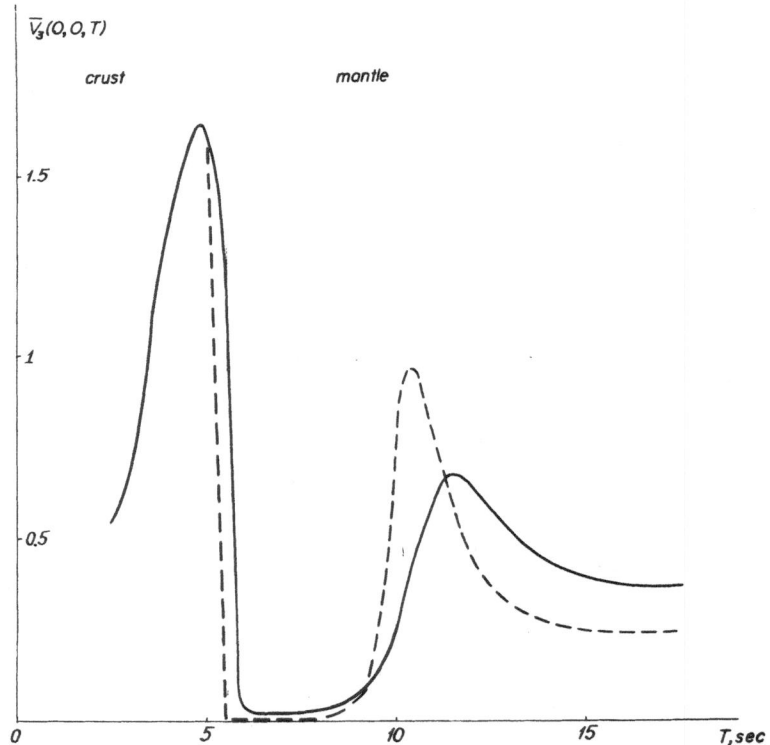

Fig. 39. Frequency characteristics of Love waves for the case of a concentrated surface source:
The continuous curves are for the spherical model, the dotted curves for the plane model.
a) Fundamental mode; b) third mode.

1. Theory

Let a small change $\delta b(l)$ in the velocity correspond to disturbances $\delta_b v_k$ and $\delta_b C_k$ in the velocities and to disturbances $\delta_b L_1$, $\delta_b L_2$, and $\delta_b L_3$ in the integrals L_i; let small variations $\delta_\rho(l)$ in the density lead to disturbances $\delta_\rho v_k$, $\delta_\rho C_k$, $\delta_\rho L_1$, etc.

It is known from the calculus of variations that the disturbance in a characteristic function $\widetilde{y}_k$ is of the second order relative to the disturbance in the corresponding characteristic ξ_k^2, and that the quantities v_k and C_k are related to this disturbance. Neglecting the disturbance in $\widetilde{y}_k$, we obtain the following relations:

$$
\left.
\begin{aligned}
\delta_\rho L_1 &= \int_{R_0}^{R} \delta\rho\,\widetilde{y}_k\,dl \\[2mm]
\delta_b L_1 &= 0 \\[2mm]
\delta_\rho L_2 &= \int_{R_0}^{R} \delta\rho\,\frac{l^2}{R^2}\,b^2\,\widetilde{y}_k^2\,dl \\[2mm]
\delta_b L_2 &= 2\int_{R_0}^{R} \delta b\,\frac{l^2}{R^2}\,b\rho\,\widetilde{y}_k\,dl \\[2mm]
\delta_\rho L_3 &= \int_{R_0}^{R} \delta\rho\,\frac{l^4}{R^4}\,b^2\,(\widetilde{y}_k')^2\,dl \\[2mm]
\delta_b L_3 &= 2\int_{R_0}^{R} \delta b\,\frac{l^4}{R^4}\,b\rho(\widetilde{y}_k')^2\,dl
\end{aligned}
\right\} \quad (14.1)
$$

$$
\left.
\begin{aligned}
\delta_\rho v_k &\approx \frac{v_k}{2L_2}\left(\delta_\rho L_2 + \frac{1}{\xi_k^2}\delta_\rho L_3 - v_k^2\,\delta_\rho L_1\right) \\[2mm]
\delta_b v_k &= \frac{v_k}{2L_2}\left(\delta_b L_2 + \frac{1}{\xi_k^2}\delta_b L_3\right) \\[2mm]
\delta_\rho C_k &= \frac{1}{v_k L_1}\delta_\rho L_2 - \frac{C_k}{L_1}\delta_\rho L_1 - \frac{C_k}{v_k}\delta_\rho v_k \\[2mm]
\delta_b C_k &= \frac{1}{v_k L_1}\delta_b L_2 - \frac{C_k}{v_k}\delta_b v_k \ .
\end{aligned}
\right\} \quad (14.2)
$$

In the program we used for the calculations of Love waves, b and ρ are piecewise-linear functions of l:

$$
b(l) = b_i + \widetilde{b}_i\,(l - l_{i-1})
$$

$$
\rho(l) = \rho_i + \widetilde{\rho}_i\,(l - l_{i-1}) \ .
$$

Here b_i is the velocity and ρ_i the density at the upper boundary of the i-th layer (i.e., for $l = l_{i-1} - 0$); $\tilde{b}_i$ and $\tilde{\rho}_i$ are the velocity gradient and density gradient in the i-th layer ($i = 1,2,\dots m$).

We also assume that the variations $\delta b(l)$, and $\delta\rho(l)$ are piecewise-linear functions:

$$\delta b (l) = \delta b_i + \delta\tilde{b}_i (l - l_{i-1}) \tag{14.3}$$

$$\delta\rho(l) = \delta\rho_i + \delta\tilde{\rho}_i (l - l_{i-1})$$

where $l_{i-1} > l > l_i$, δb_i and $\delta\rho_i$ are the disturbances in the velocity and density at the upper boundary of the i-th layer, and $\delta\tilde{b}_i$ and $\delta\tilde{\rho}_i$ are the disturbances in the gradients of these same parameters in the i-th layer. Then omitting intermediate calculations, we have

$$\delta_\rho \nu_k = \sum_{i=1}^{m}\left(\frac{\partial \nu_k}{\partial\rho_i}\ \delta\rho_i + \frac{\partial\nu_k}{\partial\tilde{\rho}_i}\ \delta\tilde{\rho}_i\right)$$

$$\tag{14.4}$$

$$\delta_b \nu_k = \sum_{i=1}^{m}\left(\frac{\partial\nu_k}{\partial b_i}\ \delta b_i + \frac{\partial\nu_k}{\partial\tilde{b}_i}\ \delta\tilde{b}_i\right)$$

and similar expressions for $\delta_\rho C_k$, and $\delta_b C_k$. Here $\partial\nu_k/\partial\rho_i$, $\partial\nu_k/\partial b_i$, $\partial\nu_k/\partial\tilde{\rho}_i$ and $\partial\nu_k/\partial\tilde{b}_i$ are the partial derivatives of the phase velocity with respect to the parameters of the i-th layer; they are given by the formulas

$$\frac{\partial\nu_k}{\partial\rho_i} = \frac{\nu_k}{2L_2}\left(J_{2i} + \frac{1}{\xi_k^2}J_{4i} - \nu_k^2 J_{1i}\right)$$

$$\frac{\partial\nu_k}{\partial b_i} = \frac{\nu_k}{2L_2}\left(J_{3i} + \frac{1}{\xi_k^2}J_{5i}\right)$$

$$\tag{14.5}$$

$$\frac{\partial\nu_k}{\partial\tilde{\rho}_i} = \frac{\nu_k}{2L_2}\left(\tilde{J}_{2i} + \frac{1}{\xi_k^2}\tilde{J}_{4i} - \nu_k^2\tilde{J}_{1i}\right).$$

$$\frac{\partial\nu_k}{\partial\tilde{b}_i} = \frac{\nu_k}{2L_2}\left(\tilde{J}_{3i} + \frac{1}{\xi_k^2}\tilde{J}_{5i}\right)$$

Here

$$J_{1i} = \int_{l_i}^{l_{i-1}}\frac{l^4}{R^4}\tilde{y}_k^2\,dl\ ; \qquad\qquad J_{2i} = \int_{l_i}^{l_{i-1}}\frac{l^2}{R^2}b^2\tilde{y}_k^2\,dl$$

$$\tilde{J}_{3i} = \int_{l_i}^{l_{i-1}}\frac{l^2}{R^2}b\rho\tilde{y}_k^2\,dl\ ; \qquad\qquad J_{4i} = \int_{l_i}^{l_{i-1}}\frac{l^4}{R^4}b^2\left(\tilde{y}_k'\right)^2\,dl \tag{14.6}$$

$$J_{5i} = 2\int_{-l_i}^{l_{i-1}}\frac{l^4}{R^4}b\rho\left(\tilde{y}_k'\right)^2\,dl\ .$$

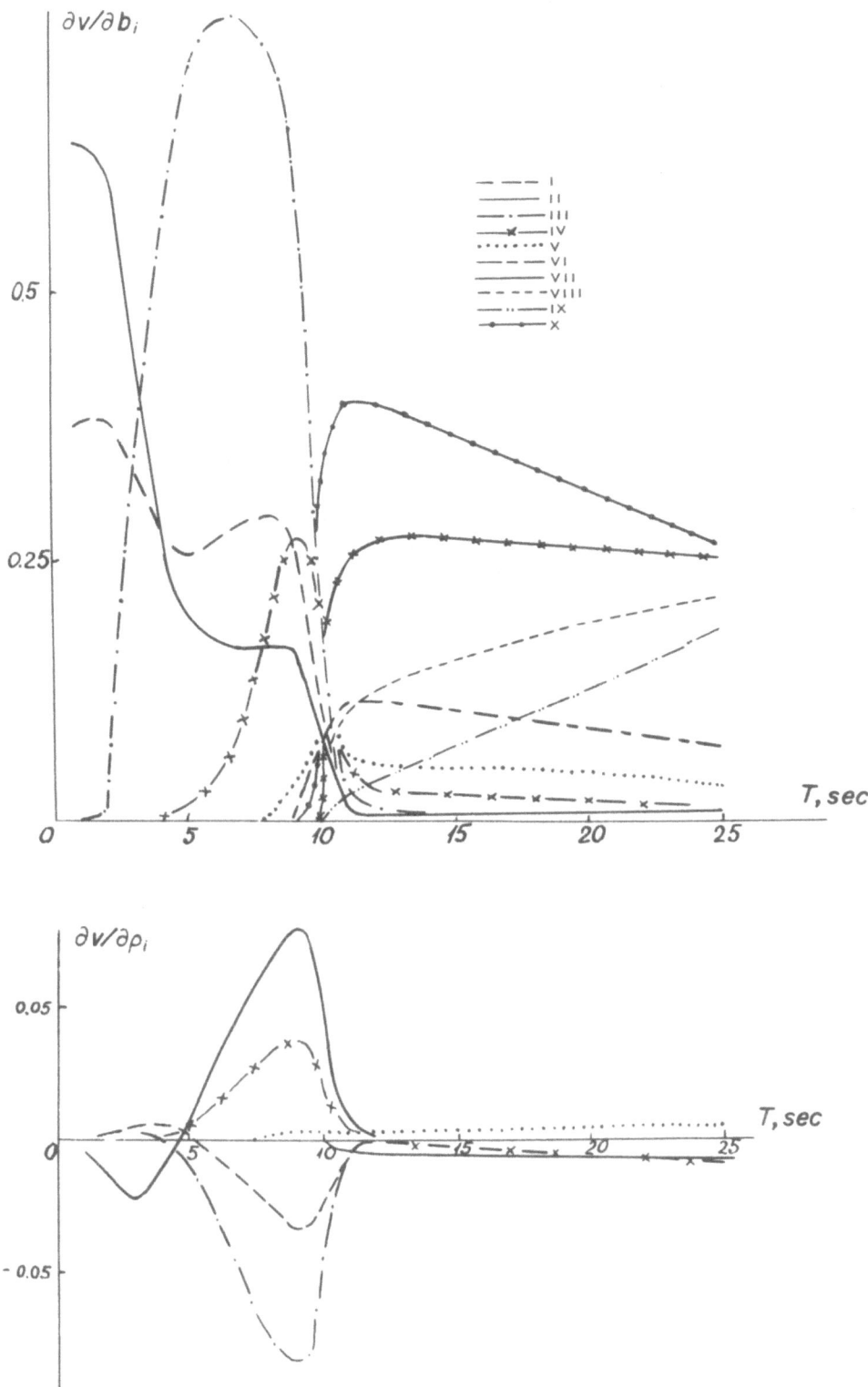

Fig. 40. Derivatives of the velocities of the second mode of Love waves with respect to the profile parameters: a) velocities b_i; b) densities ρ_i. The Roman numerals are the layer numbers. These layers are between the following depths below the surface in km: I) 0-10; II) 10-19; III) 19-38; IV) 38-60; V) 60-80; VI) 80-100; VII) 100-140; VIII) 140-170; IX) 170-200; X) 200-250.

The integrals $\tilde{J}_{1i}$ to $\tilde{J}_{5i}$ differ from the corresponding integrals J_{1i} to J_{5i} only in the extra factor $(l - l_{i-1})$ in the integrand. We obtain in exactly the same way expressions for the derivatives of the group velocity:

$$\frac{\partial C_k}{\partial \rho_i} = \frac{1}{\nu_k L_1} J_{2i} - \frac{C_k}{L_1} J_{1j} - \frac{C_k}{\nu_k} \frac{\partial \nu_k}{\partial \rho_i}$$

$$\frac{\partial C_k}{\partial b_i} = \frac{1}{\nu_k L_1} J_{3i} - \frac{C_k}{\nu_k} \frac{\partial \nu_k}{\partial b_i}$$

$$\frac{\partial C_k}{\partial \tilde{\rho}_i} = \frac{1}{\nu_k L_1} \tilde{J}_{2i} - \frac{C_k}{L_1} \tilde{J}_{1i} - \frac{C_k}{\nu_k} \frac{\partial v_k}{\partial \tilde{\rho}_k}$$

$$\frac{\partial C_k}{\partial \tilde{b}_i} = \frac{1}{v_k L_1} \tilde{J}_{3i} - \frac{C_k}{v_k} \frac{\partial v_k}{\partial \tilde{b}_i} \quad .$$

(14.7)

A special subroutine was added to the program for calculating Love waves in order to calculate the partial derivatives from (14.6)-(14.8). It operates in parallel with the subroutine for the calculation of the characteristic functions $\tilde{y}_k$. For each period T the program can output a sequence of values of the derivatives of v_k and C_k with respect to the parameters of each layer of a profile.

2. Calculations

Figure 40 contains graphs of the partial derivatives of the phase velocity of the second mode of Love waves with respect to the velocities and densities b_i and ρ_i; the velocity and density profiles are taken from the results of Gutenberg and Birch. The graphs of derivatives with respect to the gradients of $\tilde{b}_i$ and $\tilde{\rho}_i$ are similar in form; these derivatives play the usual role of small corrections to the derivatives with respect to the parameters themselves. The graphs of the derivatives of the group velocity are a little more complicated, but they are in general similar to the graphs of phase-velocity derivatives. The graphs clearly show:

1) The effect of density on the dispersion is very small compared with the effect of velocity; the effect of density on the dispersion of channel waves is particularly small (for periods of 10-20 sec).

2) The regions where the separate layers have an effect are rather sharply limited: thus for the second mode, the parameters of the crust influence the dispersion mainly in the range T < 10 sec; for T > 5-10 sec the effect of the basement layer in the roof of the mantle is also fairly strong; for T ~ 10-20 sec the dispersion is practically controlled by only the velocities in the layers forming the waveguide in the mantle at depths of 80-200 km.

The use of derivatives in calculations of the dispersion for disturbed models and in the solution of inverse problems has been investigated in several papers [57, 58]. We note that the use of derivatives of the group velocity of the fundamental mode and the use of all derivatives for the higher modes requires great care: a condition for the validity of (14.1)-(14.2) and subsequent formulas is that the disturbances in the characteristic functions be small, and this is true only in regions where the functions $v_k(T)$ and $C_k(T)$ are relatively well behaved. In zones where the graphs of these functions have sharp bends (for example at the passage of the waves from the crust into a waveguide) the characteristic functions are very unstable with respect to small disturbances in the parameters (see Section 3). Calculations show that, for disturbances $\delta b(l)$ of the order of 0.2 km/sec and higher in a layer of lowered velocity, the error in the phase velocities obtained from derivatives is of the same order as the disturbances themselves in zones where the curve for $v_k(T)$ bends sharply, i.e., they are larger than can be allowed. Hence in solving the inverse problem by using higher modes it is expedient to use the derivatives of v_k and C_k either only qualitatively (for the selection of the parameterization) or in the completion of the calculations (for verifying the accuracy of the details of a profile).

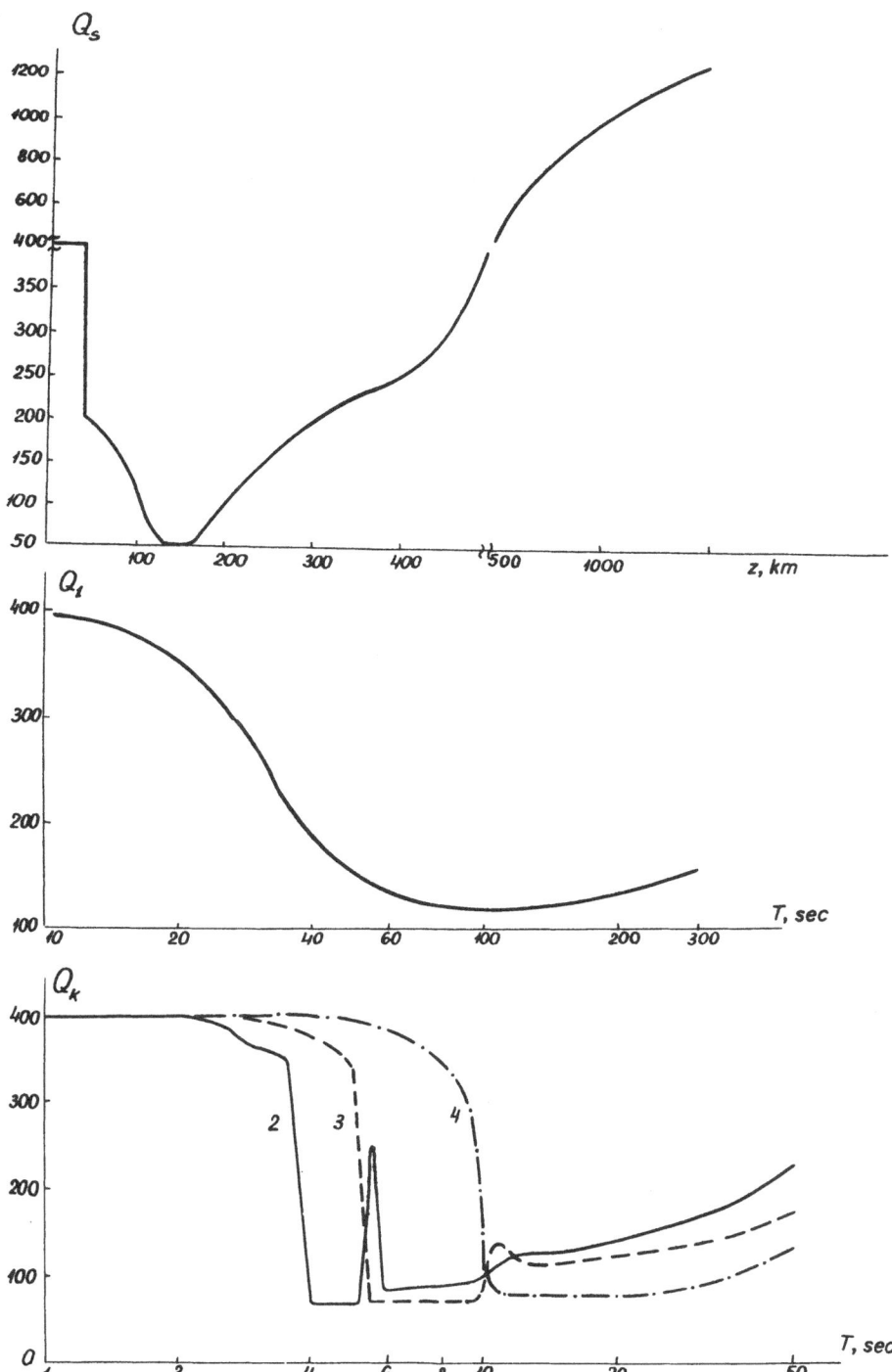

Fig. 41. The dependence of Q_k on the period in an absorbing model of the earth. a) The given distribution $Q_s(z)$; b) $Q_l(T)$; c) $Q_k(T)$, k = 2, 3, 4.

§15. The Influence of Absorption

A comparison of calculated spectra and seismograms with observed results is not very effective if the absorption of energy due to the imperfect elasticity of the medium is not taken into account. In addition to this the results of observation can be used in the solution of the inverse problem of the absorptive properties of real media. We must thus know, at least approximately, how to calculate the absorption of Love waves when the dependence of the absorption coefficient for volume waves on the frequency and the depth is known.

Imperfections in the elasticity of the medium can be described by introducing a small imaginary increment $i\delta b(l, p)$ into the transverse-wave velocity $b(l)$.* The wave number ξ_k must also have a small negative imaginary increment $-i\alpha_k$. The quantity α_k is the absorption coefficient of traveling Love waves of mode k.† The quality Q_k is related to α_k by the equation $Q_k = p_k/2\alpha_k C_k$.

Methods of calculating Q_k and α_k for a given depth distribution of $\delta b(l)$ are described in several articles [57, 59, 60]. We consider a method based on the use of the derivatives of the phase velocity.

1. Theory

Let $\delta b(l)$ satisfy (14.3) as before. Then (14.1) and (14.4) yield

$$\alpha_k = \frac{\xi_k \delta_b v_k}{v_k} = \frac{\xi_k}{v_k} \sum_{i=1}^{m} \left(\frac{\partial v_k}{\partial b_i} \delta b_i + \frac{\partial v_k}{\partial \tilde{b}_i} \delta \tilde{b}_i \right) . \tag{15.1}$$

This quantity is easily related to the local absorption coefficient $\delta b(l, p)$ or the local quality $\alpha_{SH}(l, p)$ of the SH body wave;

$$\delta b(l) = \frac{\alpha_{SH}(l, p) b^2(l)}{p} = \frac{b(l)}{2 Q_{SH}(l,p)} .$$

We write $\alpha_{SH} b^2 = q$ and $b/Q_{SH} = S$.

Let $q(l)$ and $S(l)$ be piecewise-linear functions of l (for sufficiently thin layers this corresponds approximately to the linear behavior of α_{SH} and Q_{SH}). Then

$$\alpha_k = \frac{1}{v_k^2} \sum_{k=1}^{m} \left(\frac{\partial v_k}{\partial b_i} q_i + \frac{\partial v_k}{\partial \tilde{b}_i} \tilde{q}_i \right)$$

$$Q_k = \frac{v_k^2}{C_k} \left[\sum_{k=1}^{m} \left(\frac{\partial v_k}{\partial b_i} S_i + \frac{\partial v_k}{\partial \tilde{S}_i} \tilde{S}_i \right) \right]^{-1},$$

where q_i and s_i are the values of q and s at the upper boundary of the i-th layer (for $l = l_{i-1} - 0$), and $\tilde{q}_i$ and $\tilde{S}_i$ are the gradients of these quantities in the i-th layer.

2. Calculations

Figure 41 shows values of $Q_k(T)$ for the fundamental and the three higher modes of Love waves in the models in Fig. 40. The distribution of $Q_s(l)$ is shown similarly; the minima of $Q_s(l)$ correspond to a layer of lowered velocity; it is assumed that $Q_s(l)$ is independent of the frequency.

---

*Here $\delta b(l, p)$ and $b(l)$ are real.

†The real part of the increment of ξ_k, which describes the variation of velocity caused by imperfections in the elasticity, is small (a second order quantity), and it will be omitted in the following analysis.

It is clear that Q_k depends strongly on the period in a complex fashion, even when Q_s is independent of the period. The zones of very small Q_k ($k = 2, 3, 4$) correspond to channel waves. Local zones of rapid increase of Q_k correspond to resonance phenomena in the system consisting of the crust and the layer of lowered velocity (Sections 3 and 8).

Using the same tables of the derivatives $\partial v_k / \partial b_i$ we can easily calculate $Q_k(p)$ for the case when Q_s depends on the frequency.

LITERATURE CITED

1. A. A. Abramov, A Variant of the "Progonki" Method, Zh. Vychis. Matem. i Matem. Fiz., 1, 2 (1961).*

2. V. M. Arkhangel'skaya, The Dispersion of Surface Waves in the Earth's Crust (resumé). Izv. Akad Nauk SSSR, seriya geofiz., 9. (1960).†

3. V. T. Arkhangel'skii, D. P. Kirnos, et al., Apparatus and Methods of Observation at the Seismic Stations of the USSR, Izd-vo Akad Nauk SSSR, 1962.

4. L. M. Brekhovskikh, Waves in Layered Media. Izd-vo Akad. Nauk SSSR, 1957.

5. I. Vanek, A. Zatopek, et al., Standardization of the Magnitude Scale, Izv. Akad. Nauk SSSR, seriya geofiz., 2, (1962).

6. G. Watson, The Theory of Bessel Functions. Cambridge Univ. Press, England, 1962.

7. I. M. Gel'fand and O. V. Lokutsievskii. The "Progonki" Method for the Solution of Difference Equations; S. K. Godynov and V. S. Ryaben'kii, Appendix II of Introduction to the Theory of Difference Schemes. Fizmatgiz, 1962.

8. B. Gutenberg, The Velocity of Propagation of Seismic Waves in the Earth's Crust. In: The Earth's Crust [Russian translation] IL, 1957.

9. V. I. Keilis-Borok, Surface Interference Waves. Izd-vo Akad. Nauk SSSR, 1960.

10. V. I. Keilis-Borok and T. B. Yanovskaya, The Dependence of the Spectrum of Surface waves on the Depth of the Source in the Earth's Crust. Izv. Akad. Nauk SSSR, seriya geofiz., 11 (1962).

11. E. A. Coddington and N. Levinson, The Theory of Ordinary Differential Equations. McGraw-Hill, New York, 1955.

12. B. M. Levitan, Expansions in Characteristic Functions of Second-Order Differential Equations. Gostekhizdat, Moscow-Leningrad, 1950.

13. A. L. Levshin, Love Waves in a Layer of Lowered Velocity in the Upper Mantle, Izv. Akad. Nauk SSSR, seriya geofiz., 11 (1964).

14. A. L. Levshin, The Interpretation of Data on the Dispersion of Surface Waves by Using a Bilogarithmic Grid of Theoretical Dispersion Curves, Izv. Akad. Nauk SSSR, seriya geofiz., 2 (1960).

15. M. A. Naimark, The Roots of the Frequency Equation for an Elastic Layer Lying on an Elastic Half-Space, Trudy Geofiz. In-ta Akad Nauk SSSR, 1:128 (1948).

16. C. Pekeris, The Theory of the Propagation of the Sound of Explosions in a Shallow Layer. In the Collection: The Propagation of Sound in the Ocean [Russian translation] IL, 1951.

17. G. I. Petrashen', Vibrations of an Elastic Half-Space Closed by a Layer of Liquid, Uch. Zap. Leningrad Gos. Univ. Seriya Matem. Nauk 149 (1951).

18. G. I. Petrashen', The Propagation of Elastic Waves in a Layered-Isotropic Medium Divided by Parallel Planes, Uch. Zap. Leningrad. Gos. Univ Seriya Mathem. Nauk, 149 (1951).

19. C. F. Richter, Elementary Seismology, Chapter 22, IL, 1963.

20. T. M. Sabitova, An Estimate of Focus Depths in the Earth's Crust Using Surface-Wave Spectra, Trudy Inst. Fiz., Mat. i Mekh., Akad. Nauk Kirg. SSR, 1965.

*This journal appears in cover-to-cover English translation as USSR Computational Mathematics and Mathematical Physics, published by Pergamon Press.

†This journal appears in cover-to-cover English translation as Bulletin of the Academy of Sciences of the USSR: Geophysics Series, published by American Geophysical Union.

21. S. L. Solov'ev and N. V. Shebalin, The Determination of the Intensity of Earthquakes from the Displacement of the Soil in Surface Waves, Izv. Akad. Nauk SSSR, seriya geofiz., 7 (1957).

22. C. E. Tatel and M. A. Tuve, Seismic Investigations of the Continental Crust. In: The Earth's Crust [Russian translation] Il, 1957.

23. E. C. Titchmarsh, Eigenfunction Expansions Associated with Second-Order Differential Equations, Vol. 1 Oxford Univ. Press, England, 1962.

24. V. A. Fok, Tables of Airy Functions, Moscow, 1946.

25. V. V. Khorosheva, Some Results of an Investigation of Pa and Sa Waves Using Seismograms Obtained at USSR Stations, Izv. Akad. Nauk SSSR, seriya geofiz., 7 (1960).

26. N. V. Shebalin, Magnitude and Depth of an Earthquake Focus, In: Earthquakes in the USSR, Moscow, 1961.

27. D. I. Sherman, The Distribution of Waves in a Liquid Layer on an Elastic Half-Space, Trudy Seism. Inst. Akad Nauk, 115 (1945).

28. T. B. Yanovskya, The use of Hodographs in the Calculations of Profiles of Velocities in the Upper Mantle as an Inverse Mathematical Problem, Izv. Akad. Nauk, seriya geofiz., 8 (1963).

29. T. B. Yanovskaya, The Investigation of Displacement Fields in Surface Waves with the Aim of Determining Dynamic Parameters of Earthquake Foci. Candidates Dissertation, Moscow, 1958.

30. T. B. Yanovskaya, The Determination of Dynamic Parameters of Earthquake Foci from Records of Surface Waves, Izv. Akad. Nauk SSSR, seriya geofiz., 3 (1958).

31. T. B. Yanovskaya, The investigation of Dispersive Surface Waves in the Vicinity of a Group-Velocity Minimum, Izv. Akad. Nauk SSSR, seriya geofiz., 12 (1959).

32. I. J. Azbell and T. B. Yanovskaya, The Determination of Velocities in the Upper Mantle from the Observation of P. Waves, Geophys. J., 8, 3 (1964).

33. M. Bath and A. L. Arroyo, Attenuation and Dispersion of G-Waves, J. Geoph. Res., 67, 5 (1962).

34. M. Bath, Channel Waves, J. Geoph. Res., 63, 3 (1958).

35. J. Brune, H. Benioff, and M. Ewing, Long Period Surface Waves from the Chilean Earthquake of May 22, 1960, Recorded on Linear Strain Seismographs, J. Geoph. Res., 66, 9 (1961).

36. J. Brune, J. Nafe, and J. Oliver, A Simplified Method for the Analysis and Synthesis of Dispersed Wave Trains J. Geoph. Res., 65, 1 (1960).

37. K. E. Bullen, An Introduction to the Theory of Seismology, Cambridge Univ. Press, England, 1953.

38. K. E. Bullen, Features of the Travel-Time Curves of Seismic Rays, Monthly Notices Roy. Astron. Soc., Geoph. Sup., 5, 4 (1945).

39. P. Caloi, Onde Longitudinali e Transversali Guidate dall' Astenosfera, Rend. Acad. naz. Lincei, 8, 15 (1953).

40. M. Ewing, W. Jardetzky, and F. Press, Elastic Waves in Layered Media, New York, 1957.

41. B. Gutenberg, Attenuation of Seismic Waves in the Earth's Mantle, Bull, Seism. Soc. Am., 48, 3 (1958).

42. B. Gutenberg, The Asthenosphere Low-Velocity Layer, Ann. Geofis, 12, 4 (1959).

43. J. A. Hudson, Love Waves in a Heterogeneous Medium, Geophys. J., 6, 2 (1962).

44. Z. S. Ivanova, V. I. Keilis-Borok, A. L. Levshin, and M. G. Neigaus, Love Waves and the Structure of the Upper Mantle, Geophys. J., 9, 1 (1964).

45. A. L. Levshin, M. G. Neigaus, and T. M. Sabitova, Surface Wave Spectra and the Depth of the Crust Earthquakes, Geophys J., 9, 2-3 (1964).

46. I. Lehman, The Times of P and S in Northeastern America, Ann. Geofis., 8, 4 (1955).

47. A. E. Love, Some Problems of Geodynamics, Cambridge Univ. Press, England, 1911.

48. F. Press, Some Implications on Mantle and Crustal Structure from G Waves and Love Waves, J. Geoph. Res., 64, 5 (1959).

49. F. Press and M. Ewing, Waves with Pn and Sn Velocity at Great Distances, Proc. Nat. Acad. Sci. U. S., 41, 1 (1955).

50. Y. Sato, Attenuation, Dispersion, and the Wave Guide of the G Wave, Bull, Seism. Soc. Am., 48, 3 (1958).

51. Y. Sato, M. Landisman, and M. Ewing, Love Waves in a Heterogeneous Spherical Earth, J. Geoph. Res., 65:2395-2404 (1960).

52. D. L. Anderson, and N. Toksoz, Surface Waves on a Spherical Earth J. Geoph. Res., 68, 11 (1963).

53.R. L. Kovach, and D. L. Anderson, Higher Mode Surface Waves and their Bearing on the Structure of the Earth's Mantle, Bull. Seism. Soc. Am., 54, 1 (1964).

54.R. L. Kovach and D. L.Anderson, Long-Period Love Waves in a Heterogeneous Spherical Earth, J. Geoph. Res. 67, 13 (1962).

55.Z. Alterman, H. Jarosch, and C. Pekeris, Propagation of Rayleigh Waves in the Earth, Geophys. J., 4, (1961), Jeffreys Jubilee Volume.

56.H. Jeffreys, Small Corrections in the Theory of Surface Waves, Geophys. J., 6:115-117 (1961).

57.D. L.Anderson, Universal Dispersion Tables (1), Bull, Seism. Soc. Am., 54:681-726 (1964).

58.H. Takeuchi,J. Dorman, and M. Saito, Partial Derivatives of Surface Waves Phase Velocity with Respect to Physical Parameter Changes within the Earth. J. Geoph. Res, 69, 16 (1964).

59.V. N. Zharkov, Vibration Modes of the Earth-Damping, Izv. Akad. Nauk SSSR, seriya geofiz., 2 (1962).

60.D. L. Anderson and C. B. Archambeau, Anelasticity of the Earth, J. Geoph. Res., 69, 10 (1964).

61.N. Jobert, Excitation of Torsional Oscillations of the Earth Higher Modes,J. Geophys. Res., 69, 24(1964).

62.Y. Sato, T. Usami, M. Landisman, and M. Ewing, Basic Study on the Oscillations of a Sphere (V), Geophys. J, 8, 1 (1963).